Efficient Preconditioned Solution Methods for Elliptic Partial Differential Equations

Owe Axelsson, János Karátson (editors)

Owe Axelsson, Institute of Geonics, Academy of Sciences of the Czech Republic,
Studentska 1768, 708 00, Ostrava-Poruba, Czech Republic

János Karátson, Institute of Mathematics, ELTE University,
Pf. 120, 1518 Budapest, Hungary

Contents

Chapters

Foreword

The present book is a collection of chapters written by outstanding specialists, and is devoted to one of the most challenging topics of contemporary numerical mathematics. The topic – development of efficient preconditioners and solution methods for (discretized) elliptic partial differential equations – is crucial for the mathematical modelling of complex physical and other processes in most branches of science and engineering.

Efficient solution methods usually arise as a combination of a suitable iterative technique and an efficient preconditioner, which is typically problem specific, using a proper approximation to the solved problem with further attributes as cheap actions with it, a possibility of parallel implementation, numerical and parallel scalability and robustness with respect to the problem parameters. The development of preconditioners has now its own history starting from matrix splittings, diagonal scaling and incomplete factorization for model elliptic problems and continuing with multigrid and multilevel methods, approximate inverses, domain decomposition methods and other approaches to the solution of complicated elliptic problems discretized by the finite element or other discretization methods. A substantial progress in the field of preconditioners has been achieved, but at the same time many new questions and challenges have arisen, and this is why the research is even more intensive nowadays than it was previously.

The present book covers many topics of the contemporary research in the field of efficient preconditioners, namely:

Chapter 1, written by P. Boyanova and S. Margenov, concerns the Algebraic Multilevel Iterative (AMLI) methods first introduced by Axelsson and Vassilevski. The paper overviews the methods with a special care of efficiency and robustness as well as application to both elliptic and parabolic problems discretized by either conforming or nonconforming finite element methods.

Chapter 2 by Z.-H. Cao concerns efficient preconditioning of saddle point systems, which is the timely topic appearing also in further chapters. The paper examines several preconditioners arising from either factorization or augmentation of saddle point systems with a careful examination of spectra and structure of the preconditioned matrices.

Also Chapter 3 by P. Krzyzanowski considers preconditioners for saddle point systems with an emphasis on construction of efficient block preconditioners scalable with respect to the mesh size.

Chapter 4 by K.-A. Mardal and R. Winther discusses a general approach to the construction of preconditioners, which explains scalable properties of preconditioners and can be used for the construction of new ones. It also provides examples illustrating the theory.

Chapter 5 by T. M. Austin, M. Brezina, T. A. Manteuffel and J. Ruge is devoted to preconditioning of systems arising from high order (*p*-version) FEM. A new procedure is described and tested, which constructs a sparse approximation to the FEM matrix by solving local eigenvalue and least-squares problems and then define the preconditioner by AMG approximation of the inverse of this sparse matrix.

Chapter 6 by J. Brandts, S. Korotov and M. Křížek is an exception as it is not directly devoted to the preconditioners but to tetrahedral finite element partitions. Nevertheless, these results are important for the analysis of the finite element systems, mesh refinements and also e.g. AMLI type preconditioners depending on the angle between a coarse grid space and its hierarchic completion.

Finally, Chapter 7 by R.H.W. Hoppe, X. Xu and H. Chen considers local multigrid methods applied to complicated problems from electrical engineering and acoustic.

Generally, the book provides a lot of information on recent achievements in several fields of the development of efficient preconditioners. The book appears also thanks to the publisher's understanding of the importance of such a volume and to the effort of Owe Axelsson and János Karátson, who made a selection and careful edition of the contributions. I hope that the readers appreciate this book providing an excellent overview of many new results and representing a successful joint work of authors, editors, reviewers and the publisher.

Radim Blaheta
7 November 2010
Institute of Geonics, Academy of Sciences of the Czech Republic
Studentska 1768, 708 00, Ostrava-Poruba, Czech Republic

Preface

The goal of this book is to highlight some central areas in current research on the numerical solution of elliptic problems. Mathematical models involving elliptic partial differential equations arise in a variety of real-life problems in science and engineering. Besides phenomena fully described by an elliptic equation, also time-dependent problems describing various evolutionary processes often lead to elliptic problems as subproblems arising in the course of the solution procedure. Furthermore, saddle-point problems of Stokes type, which are not elliptic in a strict sense, can be considered elliptic in a wider sense, being stationary problems that can be reduced to coercivity via the Schur complement operator. These facts reinforce the fundamental role of elliptic problems and their efficient numerical solution.

Solution methods for elliptic problems have undergone great development, and have resulted in many cases in efficient optimal algorithms, on the other hand, various new challenges induce much further research. The discretization of elliptic problems leads to algebraic systems often of very large size. To save computer memory and elapsed time, such problems are normally solved by iteration, most commonly using a preconditioned conjugate gradient (PCG) method. The proper preconditioning technique is a crucial part of the efficient solution of the arising linear systems, and this forms a major topic of this book. Particular attention is paid to multigrid and multilevel methods, and to preconditioning for saddle-point problems, often in block form. Higher order discretization methods and finite element mesh generation are also considered.

Altogether, this book provides a careful presentation of major fields in solving elliptic problems. This is done with the help of leading experts in this topic, who survey the current stage of research in individual chapters. We are convinced that both researchers in the field and users of iterative solution methods for real-life applications will benefit by these valuable contributions. We are grateful for the interest shown in this issue by the authors of these chapters. We thank also our colleague, Radim Blaheta, who due to lack of time was unable to contribute a chapter to this issue, but who kindly offered instead to write the excellent foreword to it.

Owe Axelsson
Institute of Geonics
Academy of Sciences of the Czech Republic

János Karátson
Institute of Mathematics
ELTE University, Budapest, Hungary

List of contributors

Austin, Travis

Tech-X Corporation, 5621 Arapahoe Ave, Suite A, Boulder, CO 80303
e-mail: austin@txcorp.com

Boyanova, Petia

Institute of Information and Communication Technologies, Bulgarian Academy of Sciences
Acad. G. Bonchev str, bl. 25A, 1113 Sofia, Bulgaria,
and Department of Information Technology, Uppsala University
Box 337, SE-751 05 Uppsala, Sweden; e-mail: petia@parallel.bas.bg

Brandts, Jan

Korteweg-de Vries Institute, University of Amsterdam
Science Park 904, 1098 XH Amsterdam, Netherlands; e-mail: janbrandts@gmail.com

Brezina, Marian

Department of Applied Mathematics, University of Colorado 526 UCB
Boulder, CO 80309-0526, USA; e-mail: mbrezina@math.cudenver.edu

Cao, Zhi-Hao

School of Mathematical Sciences and Laboratory of Mathematics for Nonlinear Sciences
Fudan University, Shanghai 200433, People's Republic of China; e-mail: zcao@fudan.edu.cn

Chen, Huangxin

LSEC, Institute of Computational Mathematics, Chinese Academy of Sciences
P.O.Box 2719, Beijing, 100080, People's Republic of China; e-mail: chx@lsec.cc.ac.cn

Hoppe, Ronald

Department of Mathematics, University of Houston, Houston, TX, 77204-3008, USA,
and Institute of Mathematics, University at Augsburg, D-86159 Augsburg, Germany
e-mail: rohop@math.uh.edu, hoppe@math.uni-augsburg.de

Korotov, Sergey

BCAM - Basque Center for Applied Mathematics
Bizkaia Technology Park, Building 500, E–48160 Derio, Basque Country, Spain
e-mail: korotov@bcamath.org

Křížek, Michal

Institute of Mathematics, Academy of Sciences
Žitná 25, CZ–115 67 Prague 1, Czech Republic; e-mail: krizek@math.cas.cz

Krzyżanowski, Piotr

Institute of Applied Mathematics, University of Warsaw
Banacha 2, 02-097 Warszawa, Poland; e-mail: piotr.krzyzanowski@mimuw.edu.pl

Manteuffel, Thomas A.

Department of Applied Mathematics, University of Colorado 526 UCB
Boulder, CO 80309-0526, USA; e-mail: tmanteuf@colorado.edu

Mardal, Kent-Andre

Center for Biomedical Computing, Simula Research Laboratory
P.O. Box 134, 1325 Lysaker, Norway; e-mail: kent-and@simula.no

Margenov, Svetozar

Institute of Information and Communication Technologies, Bulgarian Academy of Sciences
Acad. G. Bonchev str, bl. 25A, 1113 Sofia, Bulgaria; e-mail: margenov@parallel.bas.bg

Ruge, John

Department of Applied Mathematics, University of Colorado 526 UCB
Boulder, CO 80309-0526, USA; e-mail: jruge@colorado.edu

Xu, Xuejun

LSEC, Institute of Computational Mathematics, Chinese Academy of Sciences
P.O.Box 2719, Beijing, 100080, People's Republic of China; e-mail: xxj@lsec.cc.ac.cn.

Winther, Ragnar

Centre of Mathematics for Applications and Department of Informatics, University of Oslo
0316 Oslo, Norway; e-mail: ragnar.winther@cma.uio.no

Efficient Preconditioned Solution Methods for Elliptic Partial Differential Equations

2

Chapter 1

Robust Multilevel Preconditioning Methods

Petia Boyanova[1] and Svetozar Margenov[2]

Abstract: Preconditioners based on various multilevel extensions of two-level finite element methods (FEM) lead to iterative methods which have an optimal order computational complexity with respect to the size of the system. The methods can be based on block matrix factorized form, recursively extended via certain matrix polynomial approximations of the arising Schur complement matrices, or on additive, i.e., block diagonal form, using stabilizations of the condition number at certain levels.

In this survey, the method in focus is the Algebraic MultiLevel Iteration (AMLI). It was introduced and studied by Axelsson and Vassilevski for the case of isotropic elliptic problems in [1,2], and is referred to as AMLI. The further development of robust AMLI methods for elliptic problems is systematically discussed in the recently published monograph [3].

While a huge amount of papers are dealing with the solution of FEM elliptic systems, the related time dependent problems are much less studied. A comparative analysis of robust AMLI methods for elliptic and parabolic problems is presented in this chapter. A unified framework for both Courant conforming and Crouzeix-Raviart nonconforming linear finite element discretizations in space is used. The considered multiplicative AMLI methods are based on an approximated block two-by-two factorization of the original system matrix. A key ingredient for the efficiency of the AMLI preconditioners is the quality of the utilized block two-by-two splitting, quantified by the so-called Cauchy-Bunyakowski-Schwarz (CBS) constant, which measures the abstract angle between the two subspaces, associated with the two-by-two block splitting of the matrix. All presented methods lead to preconditioners with spectral equivalence bounds which hold uniformly with respect to both the problem and discretization parameters, and optimal total computational complexity of the related Preconditioned Conjugate Gradient (PCG) or Generalized Conjugate Gradient (GCG) solvers. The theoretical results are supported by numerical tests with an emphasis on the case of nonconforming FEM systems.

Keywords: *elliptic problems, parabolic problems, conforming FEM, nonconforming FEM, preconditioning, algebraic, multilevel, robust, CBS constant*

[1]Institute of Information and Communication Technologies, Bulgarian Academy of Sciences, Sofia, Bulgaria; e-mail: petia@parallel.bas.bg

[2](corresponding author) Institute of Information and Communication Technologies, Bulgarian Academy of Sciences, Sofia, Bulgaria; e-mail: petia@parallel.bas.bg

Owe Axelsson and János Karátson (Eds)

1   Introduction

We consider the second-order elliptic

$$
\begin{aligned}
-\nabla \cdot (\mathbf{a}(\mathbf{x})\nabla u(\mathbf{x})) &= f(\mathbf{x}) \quad \text{in} \quad \Omega, \\
u(\mathbf{x}) &= 0 \quad \text{on} \quad \Gamma_D, \\
(\mathbf{a}(\mathbf{x})\nabla u(\mathbf{x})) \cdot \mathbf{n} &= 0 \quad \text{on} \quad \Gamma_N,
\end{aligned}
\tag{1.1}
$$

and parabolic

$$
\begin{aligned}
\frac{\partial}{\partial t}u(\mathbf{x},t) - \nabla \cdot (\mathbf{a}(\mathbf{x})\nabla u(\mathbf{x},t)) &= f(\mathbf{x},t) \quad \text{in} \quad \Omega \times [0,T], \\
u(\mathbf{x},0) &= u_0 \quad \text{in} \quad \Omega, \\
u(\mathbf{x},t) &= 0 \quad \text{on} \quad \Gamma_D, \\
(\mathbf{a}(\mathbf{x})\nabla u(\mathbf{x},t)) \cdot \mathbf{n} &= 0 \quad \text{on} \quad \Gamma_N,
\end{aligned}
\tag{1.2}
$$

problems, where Ω is a polygonal domain in $\mathbb{R}^2$, $f \in L^2(\Omega)$ is a given function, $\mathbf{n}$ is the outward unit normal vector to the boundary $\partial \Omega = \overline{\Gamma}_D \cup \overline{\Gamma}_N$, $\mathbf{a}(\mathbf{x}) = \{a_{ij}(\mathbf{x})\}_{i,j \in \{1,2\}}$ is a bounded, symmetric positive definite (SPD) matrix with piecewise smooth functions $a_{ij}(\mathbf{x})$ in $\overline{\Omega} = \Omega \cup \partial \Omega$.

The weak formulation of the above problems reads as follows: for the elliptic problem (1.1), find $u \in \mathcal{V} = \{v \in H^1(\Omega) : v = 0 \text{ on } \Gamma_D\}$, such that for each $v \in \mathcal{V}$,

$$
\int_\Omega \mathbf{a}(\mathbf{x})\nabla u(\mathbf{x}) \cdot \nabla v(\mathbf{x})d\mathbf{x} = \int_\Omega f(\mathbf{x})v(\mathbf{x})d\mathbf{x},
\tag{1.3}
$$

and for the parabolic problem (1.2), find $u(\mathbf{x},t) \in \mathcal{V} \times H^1[0,T]$, $u(\mathbf{x},0) = u_0$, such that

$$
\frac{\partial}{\partial t}\int_\Omega u(\mathbf{x},t)v(\mathbf{x})d\mathbf{x} + \int_\Omega \mathbf{a}(\mathbf{x})\nabla u(\mathbf{x},t) \cdot \nabla v(\mathbf{x})d\mathbf{x} = \int_\Omega f(\mathbf{x},t)v(\mathbf{x})d\mathbf{x},
\tag{1.4}
$$

for each $v \in \mathcal{V}$.

We assume that the domain Ω is discretized by the triangulation $\mathcal{T}_\ell$ which is obtained by ℓ refinement steps of a given coarsest triangulation $\mathcal{T}_0$. We assume also that $\mathcal{T}_0$ is aligned with the discontinuities of the coefficient matrix $\mathbf{a}(\mathbf{x})$ so that on each element $T \in \mathcal{T}_0$ the entries of $\mathbf{a}(\mathbf{x}) = \mathbf{a}(e)$ are piecewise constant functions. We note that in this way big coefficient jumps across the boundaries between adjacent finite elements from $\mathcal{T}_0$ are allowed.

Remark 1.1. In the more general case where the entries of $\mathbf{a}(\mathbf{x})$ are smooth functions on each element from $\mathcal{T}_0$, we introduce an auxiliary problem with integral averaged values of $\mathbf{a}(\mathbf{x})$, i.e.,

$$
\mathbf{a}(e) = \frac{1}{|T|}\int_T \mathbf{a}(\mathbf{x})d\mathbf{x}, \qquad \forall e \subset T \in \mathcal{T}_0.
$$

Then, the related FEM matrix can be used to construct the preconditioner.

The variational elliptic problem (1.3) is then discretized using the finite element method, i.e., the continuous space $\mathcal{V}$ is replaced by a finite dimensional subspace $\mathcal{V}_h$. Then the finite element formulation is: find $u_h(\mathbf{x}) \in \mathcal{V}_h$, satisfying

$$
\sum_{e \in \mathcal{T}_h}\int_e \mathbf{a}(e)\nabla u_h(\mathbf{x}) \cdot \nabla v_h(\mathbf{x})d\mathbf{x} = \sum_{e \in \mathcal{T}_h}\int_e f(\mathbf{x})v_h(\mathbf{x})d\mathbf{x} \quad \forall v_h(\mathbf{x}) \in \mathcal{V}_h.
\tag{1.5}
$$

After discretization in space, we get the FEM elliptic system

$$K\mathbf{u} = \mathbf{f} \tag{1.6}$$

where K stands for the FEM stiffness matrix.

For the parabolic problem (1.4), we seek an approximation $u_h(\mathbf{x}, t) \in V_h$, which satisfies

$$\sum_{e \in \mathcal{T}} \frac{\partial}{\partial t} \int_e u_h(\mathbf{x}, t) v_h(\mathbf{x}) d\mathbf{x} + \sum_{e \in \mathcal{T}} \int_e \mathbf{a}(e) \nabla u_h(\mathbf{x}, t) \nabla v_h(\mathbf{x}) d\mathbf{x} = \sum_{e \in \mathcal{T}} \int_e f(\mathbf{x}, t) v_h(\mathbf{x}) d\mathbf{x}. \tag{1.7}$$

For the time discretization we use the classical θ-method,

$$M \frac{\mathbf{u}(t + \Delta t) - \mathbf{u}(t)}{\Delta t} + (1 - \theta) K \mathbf{u}(t + \Delta t) + \theta K \mathbf{u}(t) = (1 - \theta) \mathbf{f}(t + \Delta t) + \theta \mathbf{f}(t), \tag{1.8}$$

where Δt is the time step and $0 \le \theta \le 1$.

Thus, at each time step we need to solve a linear system of the form

$$(M + \Delta t (1 - \theta) K) \mathbf{u}^{n+1} = \mathbf{g}(\mathbf{u}^n), \tag{1.9}$$

for the unknown vector $\mathbf{u}^{n+1}$ at time $t + \Delta t$, where M is the FEM mass matrix and the right-hand side vector $\mathbf{g}$ depends on the approximate solution $\mathbf{u}^n$ at time t.

We study FEM discretization in space by Courant conforming and Crouzeix-Raviart nonconforming linear triangle finite elements. The related spaces $\mathcal{V}_h^C$ and $\mathcal{V}_h^{CR}$ are defined as follows.

(i) Courant conforming elements: $\mathcal{V}_h^C = \{v_h(\mathbf{x}) \in C^0(\Omega) : v_h(x)|_e \in \mathcal{P}_1(e) \, \forall e \in \mathcal{T}_h, \, v_h(\mathbf{x}) = 0 \text{ on } \Gamma_D, \}$.

(ii) Crouzeix-Raviart nonconforming elements: $\mathcal{V}_h^{CR} = \{v_h(\mathbf{x}) \in L^2(\Omega) : v_h(\mathbf{x})|_e \in \mathcal{P}_1(e) \, \forall e \in \mathcal{T}, \, v_h(\mathbf{x})$ is continuous at $\forall m_{i,e} \in \Omega, \, v_h(\mathbf{x}) = 0 \, \forall m_{i,e} \in \Gamma_D\}$, where $m_{i,e} \, (i = 1, 2, 3)$ is the midpoint of the *i*-th edge of element *e*.

The linear conforming finite elements are in general the most thoroughly studied case. However, in many applications the nonconforming elements have their strong advantages. For example, in the case of flow in highly heterogeneous porous media the finite volume and mixed finite element methods have proven to be accurate and locally mass conservative. When applying the mixed FEM, the continuity of the velocity normal to the boundary between two adjacent finite element could be enforced by Lagrange multipliers. In [4] Arnold and Brezzi have demonstrated that after the elimination of the unknowns representing the pressure and the velocity from the algebraic system, the resulting Schur system for the Lagrange multipliers is equivalent to a discretization of (1.1) by a Galerkin method using linear non-conforming finite elements. Namely, in [4] it is shown that the lowest-order Raviart-Thomas mixed finite element approximations are equivalent to the usual Crouzeix-Raviart non-conforming linear finite element approximations when the non-conforming space is augmented with cubic bubbles.

We also notice the role of the Crouzeix-Raviart elements in the recent development of iterative solution methods for discontinuous Galerkin (DG) discretizations. In [5], Ayuso de Dios and Zikatanov have considered several Interior Penalty DG methods. The uniformly convergent iterative methods they have proposed are based on a *natural* decomposition of the first order DG finite element space as a direct sum of the finite element space and a subspace that contains functions discontinuous at interior faces.

These two arguments just illustrate our motivation for the particular attention we have paid recently to the case of nonconforming FEM problems.

In this survey we provide a comparative analysis of AMLI methods for FEM systems arising after conforming and nonconforming discretization (in space) of both elliptic and parabolic problems. The rest of the chapter is organized as follows. Some basic local relations and geometric interpretations are given in the next section. Section 3 outlines the AMLI preconditioning technique and its extension to handle non-nested FEM discretizations. Section 4 is concerned with the theory of robust AMLI methods for elliptic problems while the following Section 5 is devoted to the case of parabolic problems. Some selected numerical experiments, illustrating the performance of the discussed AMLI preconditioners for both stationary and parabolic problems discretized by Crouzeix-Raviart elements, are presented and discussed in Section 6. Brief concluding remarks are given in Section 7.

2   Basic Local Relations

The domain Ω is partitioned in triangles (finite elements), and on each element we use piecewise linear approximation. The FEM subspace is spanned by a finite number of basis functions each with local support. This important property simplifies tremendously both the construction and the analysis of the local (element based) properties.

In connection with mesh refinements, one can use a partitioning of the basis functions into "previous" and "added" mesh points to form a block diagonal (additive) or Schur complement (multiplicative) preconditioner which can also be analyzed locally.

A general important technique for finite element methods is to transform the arising integral over an arbitrary element to a standard reference element. We illustrate here the transformation method for a planar domain.

The analysis for an arbitrary triangle (e) with coordinates (ξ_i, η_i), $i = 1, 2, 3$ can be done on the reference triangle $(\tilde{e}) = \{(0,0)\,(1,0)\,(0,1)\}$. Let us consider the local (element) bilinear form

$$a_e(u,v) = \int_e \mathbf{a}(e)\nabla u(\mathbf{x}) \cdot \nabla v(\mathbf{x})d\mathbf{x} = \int_e \sum_{i,j} a_{ij}\frac{\partial u}{\partial x_i}\frac{\partial v}{\partial x_j}.$$

Then the following simple relations hold true:

$$a_e(u,v) = a_{\tilde{e}}(\tilde{u},\tilde{v}) =$$

$$\int_{\tilde{e}}\left[\frac{\partial \tilde{u}}{\partial \tilde{x}_1}, \frac{\partial \tilde{u}}{\partial \tilde{x}_2}\right]\left[\begin{array}{cc}(\xi_2-\xi_1) & (\eta_2-\eta_1)\\(\xi_3-\xi_1) & (\eta_3-\eta_1)\end{array}\right]^{-1}\left[\begin{array}{cc}a_{11} & a_{12}\\a_{21} & a_{22}\end{array}\right]\left[\begin{array}{cc}(\xi_2-\xi_1) & \xi_3-\xi_1)\\(\eta_2-\eta_1) & (\eta_3-\eta_1)\end{array}\right]^{-1}\left[\frac{\partial \tilde{v}}{\partial \tilde{x}_1}, \frac{\partial \tilde{v}}{\partial \tilde{x}_2}\right]^T,$$

where

$$a_{\tilde{e}}(\tilde{u},\tilde{v}) = \int_{\tilde{e}}\sum_{i,j}\tilde{a}_{ij}\frac{\partial \tilde{u}}{\partial \tilde{x}_i}\frac{\partial \tilde{v}}{\partial \tilde{x}_j},$$

that is, $\tilde{a}_{ij}$ depend on both the shape of (e) and the coefficients a_{ij}.

Corollary 1.1. *For the local analysis of anisotropic FEM problems we can consider either the reference triangle and arbitrary coefficients $[a_{ij}]$, or alternatively, the operator $-\Delta$ and an arbitrary triangle e, see e.g., [6, 7].*

In what follows we restrict our analysis to the case of isotropic diffusion coefficients and arbitrary triangles (mesh anisotropy). Under this assumption, the next representation of the element stiffness matrix K_e holds true for both, conforming and nonconforming linear finite elements:

$$K_e = C_e \begin{bmatrix} \beta+1 & -1 & -\beta \\ -1 & \alpha+1 & -\alpha \\ -\beta & -\alpha & \alpha+\beta \end{bmatrix}, \qquad (2.1)$$

where C_e is a constant depending on the element e, $a = \cot\theta_1$, $b = \cot\theta_2$, $c = \cot\theta_3$; $\alpha = a/c$, $\beta = b/c$, $(\alpha,\beta) \in D = \{(\alpha,\beta) \in \mathbb{R}^2 : -\frac{1}{2} < \alpha \le 1, \max\{-\frac{\alpha}{\alpha+1}, |\alpha|\} \le \beta \le 1\}$.

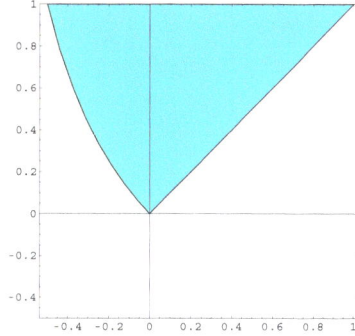

Figure (1): Domain of the parameters (α,β)

The above geometric interpretations are used, e.g., for the local analysis of the CBS constant as well as for the local construction and analysis of robust preconditioners of the first diagonal blocks which appear in hierarchical two-level and multilevel algorithms. One can find some more related details in [3, 7–13].

3   The AMLI Algorithm

The original AMLI utilizes a hierarchy of nested spaces. Since in this chapter we apply the framework also for nonconforming FEM discretizations, which imply non-nested FEM spaces, associated with consecutive nested mesh refinements, we briefly recollect the classical setting of AMLI and outline some necessary upgrades to handle non-nested spaces.
Consider a linear system of equations

$$A\mathbf{u} = \mathbf{b}.$$

The system is going to be solved via the Preconditioned Conjugate Gradient (PCG) method or, in the case of a variable preconditioner, via the Generalized Conjugate Gradient (GCG) method.
Assume that the given matrix is first split into a block two-by-two form as

$$A = \begin{bmatrix} A_{11} & A_{12} \\ A_{21} & A_{22} \end{bmatrix}. \qquad (3.1)$$

Then its exact block factorization is as follows:

$$A = \begin{bmatrix} A_{11} & 0 \\ A_{21} & S_A \end{bmatrix} \begin{bmatrix} I_1 & A_{11}^{-1}A_{12} \\ 0 & I_2 \end{bmatrix}, \qquad (3.2)$$

with the Schur complement $S_A = A_{22} - A_{21}A_{11}^{-1}A_{12}$ and I_1, I_2 being identity matrices of corresponding size. The AMLI preconditioner is based on some approximation of the exact block factorization:

$$B = \begin{bmatrix} B_{11} & 0 \\ A_{21} & S \end{bmatrix} \begin{bmatrix} I_1 & B_{11}^{-1}A_{12} \\ 0 & I_2 \end{bmatrix}. \qquad (3.3)$$

The CBS constant $0 \leq \gamma < 1$ is a measure of the splitting (see e.g. [14, 15]). However, it is clear from the factorization itself that even if we have γ very close to one, a preconditioner of the form (3.3) can be made as close to A as we want, by taking sufficiently good approximations B_{11} of A_{11}, and S of S_A. AMLI preconditioners which are (in general) independent on the splitting-related CBS constant have been recently studied. For instance, multilevel algorithms based on high quality element-by-element approximations are presented in [16–20], see also [3]. As a rule, these preconditioners are robust with respect to the coefficient jumps.

This survey has a particular focus on the robustness including the case of mesh and coefficient anisotropy. For such problems, there are still some advantages of the classical approach to find a sparse approximation of the Schur complement, using hierarchical basis functions (HBF) on a sequence of nested meshes. In the case of conforming finite elements, splitting the degrees of freedom into two parts, one corresponding to the points on a given coarse mesh and the other corresponding to the newly added points on the next finer mesh, naturally imposes a two-by-two block splitting of the related finite element matrices. Furthermore, in the HBF framework, for conforming FEM and elliptic problems, the block A_{22} turns out to be exactly the stiffness matrix associated with the coarse mesh, namely,

$$\widetilde{A}^{fine} = \begin{bmatrix} A_{11} & \widetilde{A}_{12} \\ \widetilde{A}_{21} & A^{coarse} \end{bmatrix} \begin{matrix} \}\text{fine mesh degrees of freedom only} \\ \}\text{coarse mesh degrees of freedom.} \end{matrix} \qquad (3.4)$$

Here $\widetilde{A}$ denotes the HBF matrix in contrast to A, which is the standard nodal basis matrix. It is well known that the two matrices are related via the transformation

$$\widetilde{A}^{fine} = J^T A J$$

where the transformation J has the form

$$J = \begin{bmatrix} I_1 & J_{12} \\ 0 & I_2 \end{bmatrix}. \qquad (3.5)$$

It is shown in [21, 22], see also the survey paper [15], that

$$(1 - \gamma^2)\widetilde{A}_{22} \leq \widetilde{S}_A \leq \widetilde{A}_{22}, \qquad (3.6)$$

where γ is the CBS constant, corresponding to the block two-by-two splitting (3.4). For more details on HBF methods we refer to [8, 23], and the references therein.

Remark 1.2. Throughout the chapter, the matrix relation $B \leq C$ should be understood in positive definite sense, i.e., for some positive definite matrices B and C the notation $B \leq C$ abbreviates $v^T B v \leq v^T C v$ for all nonzero vectors v of corresponding size.

Thus, provided that we can find a good approximation B_{11} to A_{11}, the two-level block-factorized preconditioner

$$\widetilde{H} = \begin{bmatrix} B_{11} & \widetilde{A}_{12} \\ \widetilde{A}_{21} & A^{coarse} \end{bmatrix} \begin{bmatrix} I_1 & B_{11}^{-1}\widetilde{A}_{12} \\ 0 & I_2 \end{bmatrix}$$

is spectrally equivalent to $\widetilde{A}$.

Given the above, the extension of the two-level construction to multilevel, combined with a certain polynomial stabilization, has led to the classical AMLI methods which possess both optimal rate of convergence and optimal computational complexity, linearly proportional to the number of degrees of freedom on the finest mesh (cf. [1, 2, 24] or also [25–28]).

We briefly sketch the AMLI framework. Consider a sequence of matrices, $A^{(\ell)}(\equiv A), A^{(\ell-1)}, \ldots, A^{(0)}$, where $A^{(k)}$ is of order n_k, which satisfies the following conditions:

- $\frac{n_{k+1}}{n_k} \geq \rho > 1$, i.e., the sizes of the matrices decrease in a geometric ratio;

- $A^{(k)}$ are sparse matrices;

- for certain sparse transformations $J^{(\ell)}, \ldots, J^{(1)}$, each matrix $\widetilde{A}^{(k)} = {J^{(k)}}^T A^{(k)} J^{(k)}$ is split into a 2×2 block form and the corresponding Schur complement is spectrally equivalent to $A^{(k-1)}$.

Then the AMLI preconditioner is defined as follows: $H^{(0)} = A^{(0)}$, and for $k = 1, \ldots, \ell$,

$$H^{(k)} = J^{(k)^{-T}} \begin{bmatrix} B_{11}^{(k)} & 0 \\ \widetilde{A}_{21}^{(k)} & [S^{(k)}] \end{bmatrix} \begin{bmatrix} I_1^{(k)} & (B_{11}^{(k)})^{-1} \widetilde{A}_{12}^{(k)} \\ 0 & I_2^{(k)} \end{bmatrix} J^{(k)^{-1}} ,$$

where $\left[S^{(k)} \right]$ denotes that certain stabilization technique is performed on some (or all) of the levels. The matrices $A^{(k)}$ may be obtained in various ways. For instance, they could be the matrices obtained by the discretization of the underlying partial differential equation on a sequences of nested meshes. For the latter case, using conforming FEM and HBF, there exist transformation matrices $J^{(\ell)}, \ldots, J^{(1)}$, of the form (3.5), such that

$$\widetilde{A}^{(k)} = {J^{(k)}}^T A^{(k)} J^{(k)} = \begin{bmatrix} A_{11}^{(k)} & \widetilde{A}_{12}^{(k)} \\ \widetilde{A}_{21}^{(k)} & A^{(k-1)} \end{bmatrix} .$$

One particular stabilization is via a matrix polynomial, namely,

$$\left[S^{(k)} \right] \equiv \widetilde{S}^{(k)} = A^{(k-1)} \left[I - P_{\nu_k}^{(k)} (H^{(k-1)^{-1}} A^{(k-1)}) \right]^{-1} . \tag{3.7}$$

In this case, the following optimality condition for $\nu_k = \nu$ holds true [2, 23], see also the monograph [3]:

$$\frac{1}{\sqrt{1-\gamma^2}} < \nu < \rho \approx 4, \tag{3.8}$$

where ρ stands for the refinement factor. This means that, if the inequalities (3.8) are satisfied, then there exists a polynomial $P_\nu^{(k)}$ such that the AMLI (3.7) has both, optimal convergence rate and optimal computational complexity.

Following [29] we present the AMLI algorithm in a *pseudo code* form:

Procedure *AMLI*: $\mathbf{u}^{(k)} \leftarrow AMLI\left(\mathbf{b}^{(k)}, k, \nu_k, \{p_j^{(k)}\}_{j=0}^{\nu_k} \right)$;

(a) $\mathbf{b}^{(k)} = {J^{(k)}}^T \mathbf{b}^{(k)},\ [\mathbf{b}_1^{(k)}, \mathbf{b}_2^{(k)}] \leftarrow \mathbf{b}^{(k)}$,

(b) $B_{11}^{(k)} \mathbf{w}_1^{(k)} = \mathbf{b}_1^{(k)}$,

(c) $\mathbf{w}_2^{(k)} = \mathbf{b}_2^{(k)} - \widetilde{A}_{21}^{(k)} \mathbf{w}_1^{(k)}$,

(d) $k = k - 1$,

(f) **if** $k = 0$ **then** $\mathbf{u}_2^{(1)} = {A^{(0)}}^{-1} \mathbf{w}_2^{(1)}$, solve on the coarsest level exactly;

(g) **else**

(h) $\qquad \mathbf{u}_2^{(k+1)} \leftarrow AMLI\left(p_{\nu_k}^{(k)}\mathbf{w}_2^{(k+1)}, k, \nu_k, \{p_j^{(k)}\}_{j=0}^{\nu_k}\right);$

(i) $\qquad$ **for** $j = 1$ **to** $\nu_k - 1$:

(j) $\qquad\qquad \mathbf{u}_2^{(k+1)} \leftarrow AMLI\left(A^{(k)}\mathbf{u}_2^{(k+1)} + p_{\nu_k-j}^{(k)}\mathbf{w}_2^{(k+1)}, k, (k)\nu_k, \{p_j^{(k)}\}_{j=0}^{\nu_k}\right);$

(l) $\qquad$ **endfor**

(m) **endif**

(n) $\quad k = k + 1,$

(o) $\quad B_{11}^{(k)}\mathbf{v}_1 = \widetilde{A}_{12}^{(k)}\mathbf{u}_2^{(k)}, \ \mathbf{u}_1^{(k)} = \mathbf{w}_1^{(k)} - \mathbf{u}_1,$

(p) $\quad \mathbf{u}^{(k)} \leftarrow [\mathbf{u}_1^{(k)}, \mathbf{u}_2^{(k)}], \ \mathbf{u}^{(k)} = J^{(k)}\mathbf{u}^{(k)}$

end Procedure *AMLI*

Remark 1.3. The stabilization in the AMLI method can be done in different ways. Besides the stabilization with a polynomial of degree ν_k (lines (h)-(l) in **Procedure** *AMLI*), an alternative option, used in Section 6, is to apply ν_k inner iterations. In the latter case, the solution process becomes nonlinear and, therefore, the Generalized CG (GCG) method, described, for instance, in [14], has to be used. One resulting method, referred to as the the I-AMLI, is derived and analyzed in [9]. In the numerical tests presented in this chapter we use nonlinear AMLI (NLAMLI) preconditioner (see e.g [3, 30]). For further information regarding stabilization of multilevel methods we refer to [24]. The particular stabilization technique, however, does not affect the theoretical considerations in this article.

In many applications of mathematical modeling in natural sciences, engineering and in other areas, the arising FEM systems are severely ill-conditioned due to some problem parameters taking values near certain limits. Examples of such parameters are ratio of coefficient jumps, coefficient and mesh anisotropy, Poisson ratio for nearly incompressible materials etc. This chapter is focused on the construction of robust AMLI methods for linear conforming and nonconforming FEM systems, which includes: a) uniform estimates of the CBS constant $\gamma < 1$; b) robust preconditioning $B_{11}^{(k)}$ of the current pivot block $\widetilde{A}_{11}^{(k)}$.

4 Robust AMLI Preconditioning of Elliptic Problems

4.1 Linear conforming FEM

Consider two consecutive meshes $\mathcal{T}_{k-1} \subset \mathcal{T}_k$. The uniform refinement of linear conforming elements is shown in Fig. (**2**). The current coarse triangle $e \in \mathcal{T}_{k-1}$ is subdivided in four congruent triangles by joining the mid-edge nodes to get the macroelement $E \in \mathcal{T}_k$. The FEM spaces corresponding to the consecutive nested triangulations are nested, and the related macroelement stiffness matrix $K_E^{(k)}$ consists of blocks which are 3×3 matrices. Therefore a local analysis is applicable to estimate the CBS constant γ (see e.g., in [3, 15]), i.e.,

$$\gamma \leq \max_{E \in \mathcal{T}_k} \gamma_E < 1. \tag{4.1}$$

The macroelement CBS constant is computed as $\gamma_E^2 = 1 - \mu_1$ where μ_1 is the minimal eigenvalue of the generalized eigenproblem

$$S_E^{(k)} \mathbf{v}_{E:2} = \mu K_e^{(k-1)} \mathbf{v}_{E:2}, \qquad \mathbf{v}_{E:2} \neq \mathbf{c}, \tag{4.2}$$

$S_E^{(k)} = K_{E:22}^{(k)} - K_{E:21}^{(k)} \left(K_{E:11}^{(k)} \right)^{-1} K_{E:12}^{(k)}$, $\mathbf{c}^T = (c, c, \cdots, c)$, c is a real constant. Therefore the local eigenproblem (4.2) has a reduced dimension of 2×2. The framework of (4.1 - 4.2) is applied to get the next theorem.

Theorem 1.1. (Maitre and Musy [13], Axelsson [8]) *For linear conforming elements and CBS constant γ_K corresponding to the stiffness matrix K, the estimate*

$$\gamma_K^2 < \frac{3}{4} \tag{4.3}$$

holds uniformly with respect to the mesh parameter, the number of refinements ℓ, coefficient jumps, and coefficient and mesh anisotropy.

The proof follows straightforwardly from the expression $\gamma_E^2 = 3/8 + 1/4\sqrt{d - 3/4}$, where $d = \sum_1^3 \cos^2 \theta_i$, see Fig. (2).

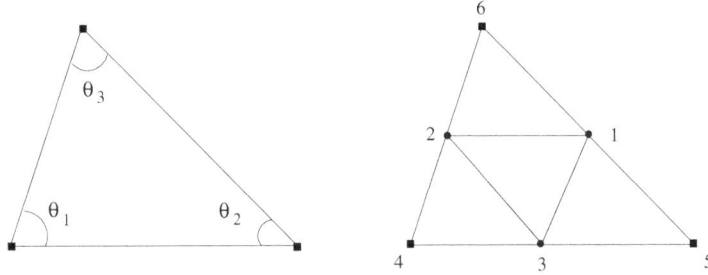

Figure (2): Uniform refinement of linear conforming finite elements

A local approach is applicable as well for preconditioning of the pivot block $K_{11}^{(k)}$. The macroelement block $K_{11:E}^{(k)}$ has the following form, see (2.1):

$$K_{11:E}^{(k)} = C_e \begin{bmatrix} \alpha + \beta + 1 & -1 & -\beta \\ -1 & \alpha + \beta + 1 & -\alpha \\ -\beta & -\alpha & \alpha + \beta + 1 \end{bmatrix}.$$

Then the additive preconditioner is $B_{11}^{(k)} = \sum_{E \in \mathcal{T}_k} B_{11:E}^{(k)}$ where the summation stands for the FEM assembling procedure, and

$$B_{11:E}^{(k)} = C_e \begin{bmatrix} \alpha + \beta + 1 & -1 & 0 \\ -1 & \alpha + \beta + 1 & 0 \\ 0 & 0 & \alpha + \beta + 1 \end{bmatrix}. \tag{4.4}$$

Theorem 1.2. (Axelsson, Padiy [9], Axelsson, Margenov [7]) *For the additive preconditioner defined by (4.4), the estimate*

$$\kappa \left(\left(B_{11}^{(k)} \right)^{-1} K_{11}^{(k)} \right) < \frac{1}{4}(11 + \sqrt{105}) \approx 5.31$$

holds uniformly with respect to the shape and size of each element and the coefficients in the differential operator.

In [7], an alternative multiplicative preconditioner of the pivot block was proposed, improving the uniform robust estimate from Theorem 1.2.

Theorem 1.3. (Axelsson, Margenov [7]) *For the multiplicative preconditioner, the estimate*

$$\kappa\left(\left(B_{11}^{(k)}\right)^{-1}K_{11}\right) < \frac{15}{8} = 1.875$$

holds uniformly with respect to the shape and size of each element and the coefficients in the differential operator.

Remark 1.4. It is interesting to note that the proof of both Theorem 1.2 and Theorem 1.3 is based on the local properties of the element stiffness matrix and on the pure algebraic inequality

$$\frac{\alpha\beta + \alpha + \beta + 1}{(\alpha + \beta + 1)(\alpha + \beta + 2)} > \frac{4}{15} \tag{4.5}$$

which holds for all $(\alpha, \beta) \in D$, see Fig. **(1)**.

In order to have optimal total computational complexity of the AMLI algorithm, the systems with the preconditioners of the pivot blocks have to be solved with optimal complexity. This can be achieved for both, the additive and the multiplicative preconditioners of the pivot block. Here we briefly discuss the additive case. We see in Fig. **(3)** that the coupled nodes of the additive preconditioner form either

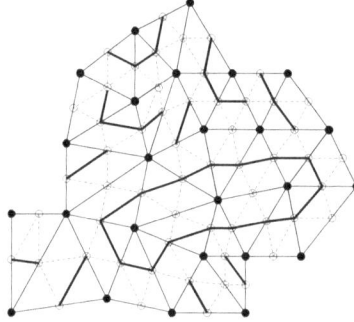

Figure **(3)**: An example of the connectivity pattern for the additive preconditioner for a mesh after one refinement step of $\mathcal{T}_0$, i.e., for $\mathcal{T}_1$

a single point, a polyline or a polygon. Therefore, there are no cross-points, the block matrix $B_{11}^{(k)}$ has a generalized tridiagonal structure, and the direct solver has optimal computational complexity. Combining the results from Theorem 1.1, Theorem 1.2 and Theorem 1.3, we obtain a class of optimal AMLI algorithms satisfying the optimality condition (3.8) for $\nu \in \{2, 3\}$. It is important to note that all of them are robust with respect to mesh and coefficient anisotropy.

4.2 Linear nonconforming FEM

The uniform refinement of linear nonconforming elements is shown in Fig. **(4)**. As seen before, for the nonconforming Crouzeix-Raviart finite element, where the nodal basis functions are defined at the midpoints along the edges of the triangle, the natural vector spaces $\mathcal{V}_{k-1}(E) := span\{\phi_I, \phi_{II}, \phi_{III}\}$ and $\mathcal{V}_k(E) := span\{\phi_i\}_{i=1}^9$ (cf. the macroelement in Fig.**(4)**(b)) are no longer nested, i.e. $\mathcal{V}_{k-1}(E) \not\subseteq \mathcal{V}_k(E)$. This means that a proper hierarchical basis is to be used.

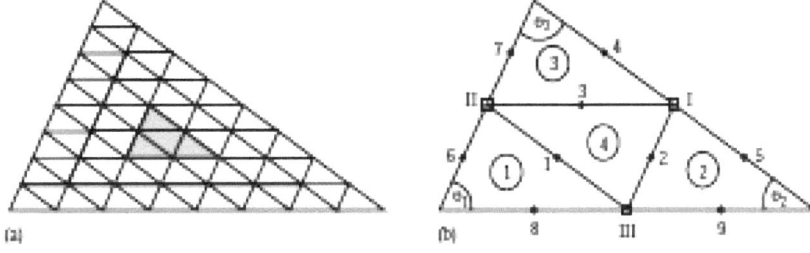

Figure (**4**): Crouzeix-Raviart finite element: (a) Discretization. (b) Macro-element

Here, we will consider in a unified setting the so-called First Reduce (FR) and Differences and Aggregates (DA) splittings [10, 12]. Following the nodes' numbering from Fig. (**4**)(b), the macroelement transformation matrix corresponding to this general splitting is given by

$$J_E = J_E(C_E) = \begin{bmatrix} I & 0 & C_E \\ 0 & J_{E-} & J_{E+} \end{bmatrix} \quad (\in \mathbb{R}^{9 \times 9}), \tag{4.6}$$

where I denotes the 3×3 identity matrix, and C_E is a 3×3 matrix whose entries c_{ij} will be specified later. The 3×6 matrices

$$J_{E-} = \frac{1}{2} \begin{bmatrix} 1 & -1 & & & & \\ & & 1 & -1 & & \\ & & & & 1 & -1 \end{bmatrix}^T, \quad J_{E+} = \frac{1}{2} \begin{bmatrix} 1 & 1 & & & & \\ & & 1 & 1 & & \\ & & & & 1 & 1 \end{bmatrix}^T \tag{4.7}$$

introduce the so-called half-difference and half-sum basis functions associated with the sides of the macro-element triangle. The global transformation matrix associated with the locally introduced J_E is denoted by J.

We consider also the transformation matrix

$$J_\pm := \begin{bmatrix} I & 0 & 0 \\ 0 & J_- & J_+ \end{bmatrix}, \tag{4.8}$$

where the global matrices J_- and J_+ correspond to the macroelement terms as introduced in (4.7), and let

$$\bar{K}^{(k)} := J_\pm^T K^{(k)} J_\pm = \begin{bmatrix} \bar{K}_{11}^{(k)} & \bar{K}_{12}^{(k)} & \bar{K}_{13}^{(k)} \\ \bar{K}_{21}^{(k)} & \bar{K}_{22}^{(k)} & \bar{K}_{23}^{(k)} \\ \bar{K}_{31}^{(k)} & \bar{K}_{32}^{(k)} & \bar{K}_{33}^{(k)} \end{bmatrix} \tag{4.9}$$

denote the matrix after the transformation step, where $\bar{K}_{11}^{(k)}$ corresponds to the interior nodes with respect to the macroelements $E \in \mathcal{T}_k$, and $\bar{K}_{22}^{(k)}$ and $\bar{K}_{33}^{(k)}$ correspond to the half-difference and half-sum basis functions introduced by (4.7). Then in the FR decomposition, the unknowns related to the pivot block $\bar{K}_{11}^{(k)} (= K_{11}^{(k)})$ are first eliminated. Let us note that $K_{11}^{(k)}$ has a block-diagonal structure, with diagonal blocks corresponding to each of the macroelements, i.e., the elimination (static condensation) can be performed locally.

Definition 1.1. *(First Reduce (FR) splitting)* The splitting based on differences and aggregates incorporating the "first reduce" step (in short FR splitting), cf. [12], is characterized by using $C_E = -\left(K_{E:11}^{(k)}\right)^{-1} \bar{K}_{E:13}^{(k)}$ in the general transformation matrix (4.6). The matrices $\bar{K}_{E:11}^{(k)} = K_{E:11}^{(k)}$ and $\bar{K}_{E:13}^{(k)}$ correspond to (4.9).

Definition 1.2. *(Differences and Aggregates (DA))* The basis transformation for the standard splitting based on differences and aggregates (DA), cf. [10], follows from the general transformation (4.6) for the choice $C_E = \frac{1}{2}\operatorname{diag}(1,1,1)$.

Then we write both the DA and FR hierarchical stiffness matrices in a 2×2 block form

$$\tilde{K}^{(k)} := J^T K^{(k)} J = \begin{bmatrix} \tilde{K}_{11}^{(k)} & \tilde{K}_{12}^{(k)} \\ \tilde{K}_{21}^{(k)} & \tilde{K}_{22}^{(k)} \end{bmatrix}.$$

These splittings utilize the AMLI recursive procedure where the aggregated block $\tilde{K}_{22}^{(k)}$ (corresponding to the half-sum basis functions) is associated with the coarser grid stiffness matrix.

Similarly to the case of conforming elements, the next two theorems provide robust estimates ensuring the optimality of the related AMLI algorithms for nonconforming elements. The condition (3.8) is then satisfied for (some) $\nu \in \{2,3\}$.

Theorem 1.4. (Blaheta, Margenov, Neytcheva [10], Kraus, Margenov, Synka [12]) *Let us denote by* $\gamma_{FR:K}$ *and* $\gamma_{DA:K}$ *the CBS constants corresponding to the above defined FR and DA hierarchical splittings of the Crouzeix-Raviart stiffness matrix. Then, the estimate*

$$\gamma_{FR:K}^2 \leq \gamma_{DA:K}^2 = \frac{3}{4}$$

holds uniformly with respect to coefficient and mesh anisotropy.

Let us note that the considered AMLI method (see (3.7)) in case of DA splitting requires stabilization polynomial of degree $\nu = 3$ while for FR it is enough to use $\nu = 2$.

Theorem 1.5. (Blaheta, Margenov, Neytcheva [11]) *For the additive preconditioners of* $\tilde{K}_{11}^{(k)}$ *blocks in the case of Crouzeix-Raviart finite elements, the following estimates hold uniformly with respect to mesh and coefficient anisotropy:*

a) additive algorithm: $\kappa\left(\left(B_{11}^{(k)}\right)^{-1}\tilde{K}_{11}\right) < \frac{1}{4}(11 + \sqrt{105}) \approx 5.31,$

b) multiplicative algorithm: $\kappa\left(\left(B_{11}^{(k)}\right)^{-1}\tilde{K}_{11}\right) < \frac{15}{8} = 1.875$

The proof of the above theorems is based again on the local relations and geometric interpretations from Section 2 as well as the inequality (4.5).

5 Robust AMLI Preconditioning of Parabolic Problems

The comparative analysis presented in this section starts with some related results published in [13,29]. The study of the parabolic problem is reduced to a matrix in the form

$$A^{(k)} = \zeta M^{(k)} + K^{(k)} \tag{5.1}$$

where $\zeta > 0$, and $M^{(k)}$ and $K^{(k)}$ stand for the FEM mass and stiffness matrices corresponding to the triangulation $\mathcal{T}_k$.

Theorem 1.6. (Maitre, Musy [13]) *Let us consider the case of linear conforming finite elements. The macroelement CBS constant satisfies*

$$\gamma_{M:E}^2 = \frac{9}{10},$$
(5.2)

then

$$\gamma_A^2 \le \gamma_{A:E}^2 \le \max\left\{\frac{3}{4}, \frac{9}{10}\right\},$$

and therefore

$$\gamma_A^2 \le \frac{9}{10}.$$
(5.3)

The estimates are uniform with respect to mesh and coefficient anisotropy.

The proof of the theorem follows from (5.2) which is straightforwardly computed by the explicit presentation of the element mass matrix

$$M_e^{(k)} = \frac{S}{12}\begin{bmatrix} 2 & 1 & 1 \\ 1 & 2 & 1 \\ 1 & 1 & 2 \end{bmatrix}$$

where $S = |e|$, $e \in \mathcal{T}_k$.

Remark 1.5. The AMLI optimality condition (3.8) is not satisfied if $\gamma^2 > 8/9$. As seen from (5.3), this could happen for certain values of the factor $\zeta > 0$ in (5.1), which is proportional to $1/\Delta t$.

Similar results for nonconforming elements were recently obtained in [29].

Theorem 1.7. (Boyanova, Margenov, Neytcheva [29]) *Let us consider the case of linear nonconforming finite elements and the related DA and FR hierarchical splittings.*

(i) DA algorithm:

The macroelement CBS constant satisfies

$$\gamma_{M:E}^2 = \frac{1}{2},$$
(5.4)

then

$$\gamma_{DA:A}^2 \le \gamma_{A:E}^2 \le \max\left\{\frac{3}{4}, \frac{1}{2}\right\},$$
(5.5)

and therefore

$$\gamma_{DA:A}^2 \le \frac{3}{4}.$$

(ii) FR algorithm:

The macroelement CBS constants are uniformly bounded with respect to mesh and coefficient anisotropy satisfying the relations

$$\gamma_{M:E}^2 = 0, \qquad \gamma_{K:E}^2 < \frac{3}{4}.$$
(5.6)

Let us note that in this case the mass matrix is diagonal and

$$M_e^{(k)} = \frac{S}{3}\begin{bmatrix} 1 & 0 & 0 \\ 0 & 1 & 0 \\ 0 & 0 & 1 \end{bmatrix}.$$

Remark 1.6. For DA, the AMLI optimality condition (3.8) for the composed matrix $A = \zeta M + K$ is satisfied for $\nu = 3$. However, in many cases the stabilization is achieved for $\nu = 2$ as well, both for DA and FR.

A numerical study of the behavior of $\gamma_E = \gamma_{E:A}$ for $A = \zeta M + K$ as a function of the factor ζS and the mesh anisotropy is shown in Figures (5)–(7). Let us denote by γ_E^C, $\gamma_{E:DA}^{NC}$ and $\gamma_{E:FR}^{NC}$ the macroelement CBS constants corresponding respectively to the case of linear conforming elements, and the cases of DA and FR splittings for linear nonconforming elements. The anisotropy is represented by the

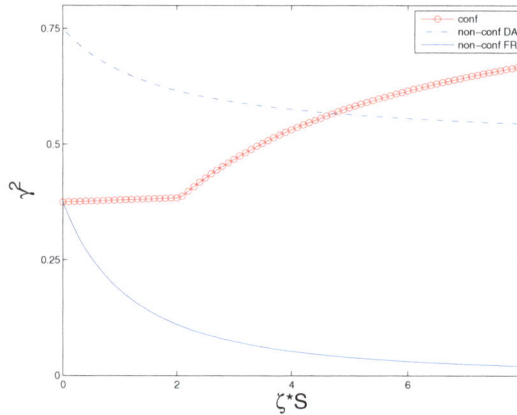

Figure (5): Triangle macroelement: $(\theta_1, \theta_2, \theta_3) = (60^o, 60^o, 60^o)$

considered three cases of triangle macroelements E, determined by their angles $(\theta_1, \theta_2, \theta_3)$. The CBS

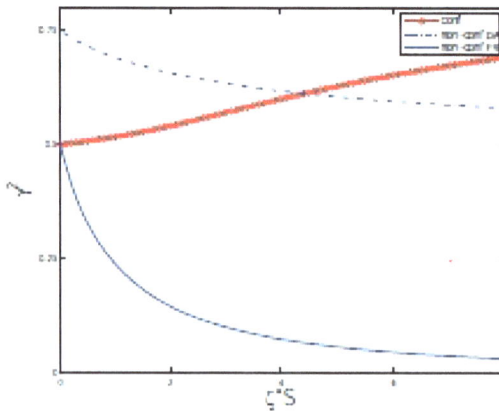

Figure (6): Triangle macroelement: $(\theta_1, \theta_2, \theta_3) = (45^o, 45^o, 90^o)$

constant of the nonconforming FR splitting, i.e., $\gamma_{E:FR}^{NC}$ is always smaller. When we compare γ_E^C and $\gamma_{E:DA}^{NC}$, we can see some advantages of the conforming case for smaller values of the factor ζS, and for cases of weaker anisotropy. The general observation is that the advantages (in this respect) of the nonconforming elements are well expressed in the last case of strongest mesh anisotropy.

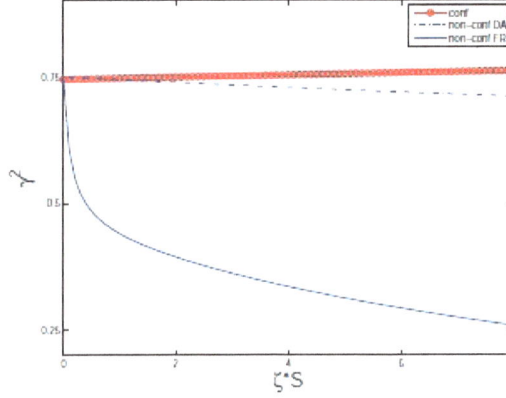

Figure **(7)**: Triangle macroelement: $(\theta_1, \theta_2, \theta_3) = (6^o, 6^o, 168^o)$

6   Numerical Tests

The numerical tests illustrate the robustness of the considered AMLI algorithms. They cover both the cases of elliptic and parabolic problems. The number of PCG/GCG iterations is given as a measure for computational efficiency. Following the discussed theoretical results, we have restricted our presentation to the case of Crouzeix-Raviart nonconforming elements.

Similar numerical tests (including isotropic problems like TP2, defined bellow) for conforming elements can be found, e.g., in [16, 17, 19]. There, the AMLI convergence, independent of the value of the splitting-related CBS constant, is based on a high quality element-by-element approximation of the Schur complement. The results are robust with respect to the coefficient jumps.

In the results presented here the pivot blocks $\tilde{A}_{11}^{(k)}$ or $\tilde{K}_{11}^{(k)}$ are preconditioned using the robust additive algorithm as defined by (4.4). This ensures an optimal order total computational complexity of the method. To improve the efficiency we implement several (up to three) inner PCG iterations to handle a better approximation of the solution of the systems with block matrices $\tilde{A}_{11}^{(k)}$ or $\tilde{K}_{11}^{(k)}$. This leads to a nonlinear AMLI preconditioner where the Generalized CG (GCG) method is used as an outer iterative scheme. One could find more details related to nonlinear AMLI in [9,23,30], or in the monograph [3]. Our particular implementation is based on the NLAMLI algorithm from [30], see also [29], where v GCG iterations are recursively performed instead of using a stabilization polynomial of degree v.

Most of the results are presented in a table form where V stands for the so called V-cycle AMLI where no stabilization is used (and PCG is applied), and W2 and W3 denote the cases of NLAMLI with $v = 2$ and $v = 3$ respectively. Two test problems denoted by TP1 and TP2 are presented.

TP1: The system $(M + \Delta t(1 - \theta)K)\mathbf{u} = \mathbf{g}$ (or $K\mathbf{u} = \mathbf{g}$ for the stationary case) is considered on a triangular domain Ω defined by its angles. The influence of the mesh anisotropy is tested by the following cases:
(a) $\theta_1 = 90^o$, $\theta_2 = \theta_3 = 45^o$,
(b) $\theta_1 = 156^o$, $\theta_2 = \theta_3 = 12^o$,
(c) $\theta_1 = 177^o$, $\theta_2 = 2^o$, $\theta_3 = 1^o$.
The coordinates of the vertices of Ω corresponding to θ_1 and θ_2 are fixed to (0,0) and (0,1) respectively.

TP2: The heat equation in $\Omega = [0, 1] \times [0, 1]$ with a discontinuous initial condition is considered. The initial condition and the numerical solutions for certain times of the evolution process are shown in

Table 1: Number of PCG/GCG iterations for the stationary problem

ℓ	TP1 (a)			TP1 (b)			TP1 (c)		
	V	W2	W3	V	W2	W3	V	W2	W3
	DA variant of the preconditioner								
1	8	11	11	10	14	14	7	10	9
2	12	13	12	19	15	14	14	18	17
3	15	13	12	34	16	14	26	22	17
4	19	13	11	62	18	14	51	26	16
5	26	13	12	114	18	14	102	25	16
	FR variant of the preconditioner								
1	8	8	8	9	9	9	6	7	6
2	11	8	8	15	10	9	10	7	6
3	14	8	8	20	10	9	16	7	6
4	16	8	7	25	9	9	21	6	6
5	20	8	8	29	9	9	19	6	6

Table 2: Number of PCG/GCG iterations for the parabolic problem: $\theta = 0$, $\Delta t = h^2$

ℓ	TP1 (a)			TP1 (b)			TP1 (c)		
	V	W2	W3	V	W2	W3	V	W2	W3
	DA variant of the preconditioner								
1	7	7	7	8	8	8	5	5	5
2	8	7	7	10	8	8	7	6	5
3	11	7	7	13	8	8	10	6	5
4	14	7	7	15	8	8	10	6	5
5	19	7	7	18	8	8	12	6	5
	FR variant of the preconditioner								
1	6	6	6	8	8	8	5	6	6
2	7	6	6	10	8	8	7	6	5
3	7	6	6	9	8	8	8	5	5
4	7	6	6	9	8	8	9	5	5
5	7	6	6	10	8	8	9	5	5

Table **3**: Averaged number of PCG/GCG iterations and reduction factors for TP2

n_ℓ	ℓ	V		W2	
		Av. number of iterations	Av.reduction factor	Av. number of iterations	Av.reduction factor
		FR variant of the preconditioner: $\theta = 0.5$; $\Delta t = h$			
800	1	6	0.261	5	0.212
3136	2	8	0.445	6	0.300
12416	3	11	0.559	6	0.349
49408	4	13	0.648	7	0.424
197120	5	16	0.734	8	0.506

Figure **(8)**.

The iteration counts for corresponding number of refinements ℓ of the initial triangulation are presented in Tables **1–3**. The presented numerical tests are in a very good agreement with the theoretical results. The number of AMLI iterations is stabilized robustly with respect to anisotropy. Many other related numerical results are currently available in, e.g., [9, 12, 29] as well as in the monograph [3].

7   Concluding Remarks

In this chapter we have summarized the theory of robust optimal order HBF-based AMLI methods to precondition matrices arising from discretization of elliptic and parabolic problems discretized by conforming and nonconforming Crouzeix-Raviart FEM.

In the case of nonconforming FEM, the so-called 'differences and aggregates' (DA) and 'first reduce' (FR) approaches are used to define hierarchical splittings of the arising non-nested FEM spaces. The DA splitting has some advantages regarding the completeness of the theoretical results. However, all known estimates for FR algorithms are always better.

Here we present numerical tests for FR based multilevel preconditioners as well. The results fully confirm the advantages of this approach in the NLAMLI setting, especially for the stationary problem in cases of strong anisotropy.

Both the theoretical and numerical results show that the proposed preconditioning technique yields a robust AMLI method for both elliptic and parabolic FEM systems.

A rather common expectation is that a good preconditioner for elliptic problems will work even better in the case of parabolic problems due to the good conditioning of the mass matrix. We have seen that this does not hold in the case of AMLI preconditioning of systems arising after a linear conforming FEM discretization.

Figure (8): Crouzeix-Raviart numerical solution of TP2

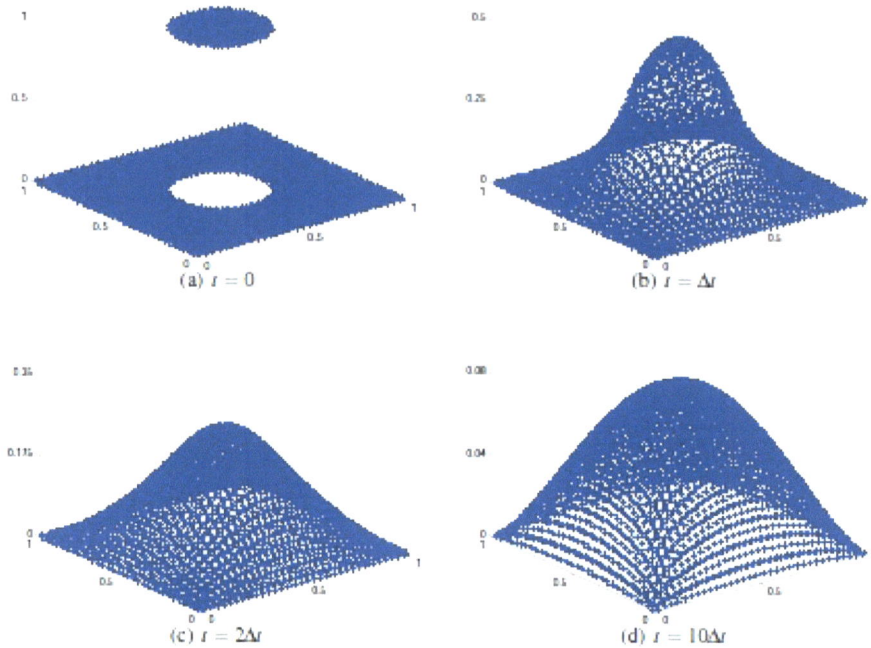

(a) $t = 0$

(b) $t = \Delta t$

(c) $t = 2\Delta t$

(d) $t = 10\Delta t$

Acknowledgments

First of all, we would like to thank the editors for the valuable remarks and suggestions concerning this chapter.

We would also like to thank Maya Neytcheva for the joint work on the parabolic problems and for the software modules provided.

This work has been partly supported by the Bulgarian NSF Grants DO 02-115/08 and DO 02-338/08. The first author is also partially funded by a grant from the Swedish Research Council (VR), *Finite element preconditioners for algebraic problems as arising in modelling of multiphase microstructures*, 2009-2011.

Some of the experiments presented in this chapter were performed on the servers of Uppsala Multidisciplinary Center for Advanced Computational Science (UPPMAX). We would like to thank UPPMAX for the access provided.

Bibliography

[1]   Axelsson O, Vassilevski PS. Algebraic multilevel preconditioning methods I. Numer Math 1989; 56: 157–177.

[2]   Axelsson O, Vassilevski PS. Algebraic multilevel preconditioning methods II. SIAM J Numer Anal 1990; 27: 1569–1590.

[3]   Kraus J, Margenov S. Robust Algebraic Multilevel Methods and Algorithms, Radon Series on Computational and Applied Mathematics 5. Berlin-New York: Walter de Gruyter 2009.

[4]   Arnold DN, Brezzi F. Mixed and nonconforming finite element methods: Implementation, postprocessing and error estimates. RAIRO Model Math Anal Numer 1985; 19: 7–32.

[5]   Ayuso de Dios B, Zikatanov L, Uniformly convergent iterative methods for discontinuous Galerkin discretizations. SIAM J Sci Comput 2009; 40: 4-36.

[6]   Axelsson O, Blaheta R. Two simple derivations of universal bounds for the CBS inequality constant. Appl Math 2004; 49(1): 57-72.

[7]   Axelsson O, Margenov S. On multilevel preconditioners which are optimal with respect to both problem and discretization parameters. Computational Methods in Applied Mathematics 2003; 3(1): 6–22.

[8]   Axelsson O. Stabilization of algebraic multilevel iteration methods; additive methods. Numerical Algorithms 1999; 21: 23–47.

[9]   Axelsson O, Padiy A. On the additive version of the algebraic multilevel iteration method for anisotropic elliptic problems. SIAM J Sci Comput 1999; 20: 1807–1830.

[10] Blaheta R, Margenov S, Neytcheva M. Uniform estimate of the constant in the strengthened CBS inequality for anisotropic non-conforming FEM systems. Num Lin Alg Appl 2004; 11: 309–326.

[11] Blaheta R, Margenov S, Neytcheva M. Robust optimal multilevel preconditioners for non-conforming finite element systems. Num Lin Alg Appl 2005; 12: 495–514.

[12] Kraus J, Margenov S, Synka J. On the multilevel preconditioning of CrouzeixÂ-Raviart elliptic problems. Num Lin Alg Appl 2008; 15: 395–416.

[13] Maitre JF, Musy F. The contraction number of a class of two-level methods; an exact evaluation for some finite element subspaces and model problems. Lect Notes Math 1982; 960: 535–54.

[14] Axelsson O. Iterative Solution Methods. Cambridge University Press 1994.

[15] Eijkhout V, Vassilevski PS. The role of the strengthened Cauchy-Bunyakowski-Schwarz inequality in multilevel methods. SIAM Review 1991; 33: 405–419.

[16] Axelsson O, Blaheta R, Neytcheva M. A black-box generalized conjugate gradient minimum residual method based on variable preconditioners and local element approximations. Technical Report No.: 2007-033. Institute for Information Technology, Uppsala University 2007.

[17] Axelsson O, Blaheta R, Neytcheva M. Preconditioning of boundary value problems using elementwise Schur complements. SIAM J Matrix Anal Appl 2009; 31(2): 767-Â789.

[18] Kraus J. Algebraic multilevel preconditioning of finite element matrices using local Schur complements. Num Lin Alg Appl 2006; 13: 49–70.

[19] Neytcheva M. On element-by-element Schur complement approximations. Linear Algebra and its Applications 2010, doi:10.1016/j.laa.2010.03.031.

[20] Bängtsson E, Neytcheva M. An agglomerate multilevel preconditioner for linear isostasy saddle point problems. Springer LNCS 2006; 3743: 113–120.

[21] Bank R, Dupont T. An optimal order process for solving finite element equations. Math Comp 1981; 36: 427–458.

[22] Axelsson O, Gustafsson I. Preconditioning and two-level multigrid methods of arbitrary degree of approximations. Math Comp 1983; 40: 219–242.

[23] Axelsson O, Vassilevski PS. A black box generalized conjugate gradient solver with inner iterations and variable-step preconditioning. SIAM J Matrix Anal Appl 1991; 12: 625–644.

[24] Vassilevski PS. Multilevel Block Factorization Preconditioners. Matrix-based analysis and algorithms for solving finite element equations. New York: Springer 2008.

[25] Axelsson O, Neytcheva M. Algebraic multilevel iterations for Stieltjes matrices. Num Lin Alg Appl 1994; 1: 213–236.

[26] Lazarov RD, Vassilevski PS, Margenov S. Solving elliptic problems by the domain decomposition method using precondition matrices derived by multilevel splitting of finite element matrix. In: Proceedings of the 1st Int. Conf. on Supercomputing; 1987; Greece; pp. 826–835.

[27] Vassilevski PS. Hybrid V-cycle algebraic multilevel preconditioners. Math Comp 1992; 58: 489–512.

[28] Vassilevski PS. On two ways of stabilizing the hierarchical basis multilevel methods. Siam Review 1997; 39: 18–53.

[29] Boyanova P, Margenov S, Neytcheva M. Robust AMLI methods for time-dependent FEM systems. J Comp Appl Math 2010; 235(2): 380–390.

[30] Kraus J. An algebraic preconditioning method for M-matrices: Linear versus nonlinear multilevel iteration. Num Lin Alg Appl 2002; 9: 599–618.

Chapter 2

Efficient Preconditioners for Saddle Point Systems

Zhi-Hao Cao[1]

Abstract: Various block preconditioners for two by two block linear saddle point systems are studied. All block preconditioners are derived from a splitting of the (1,1) block of the two by two block matrix. We analyze the properties of the corresponding preconditioned matrices, in particular their spectra, and discuss the computational performance of the preconditioned iterative methods. It is shown that fast convergence depends mainly on the quality of the splitting of the (1,1) block. Moreover, some strategies of the implementation of the block preconditioners based on purely algebraic considerations are discussed. Thus, applying our block preconditioners to the related saddle point problems, we obtain preconditioned iterative methods in a "black box" fashion.

Keywords: *saddle point systems, iterative methods, preconditioning, convergence, algebraic, eigenvalues, block preconditioners*

1   Introduction

The subject of this chapter is the solution of block 2×2 linear systems of the form

$$\mathcal{A}\begin{bmatrix} x \\ y \end{bmatrix} \equiv \begin{bmatrix} A & B^T \\ C & -D \end{bmatrix}\begin{bmatrix} x \\ y \end{bmatrix} = \begin{bmatrix} f \\ g \end{bmatrix}, \tag{2.1.1}$$

where $A \in \mathbb{R}^{n \times n}, B, C \in \mathbb{R}^{m \times n}, D \in \mathbb{R}^{m \times m}$ with $n \geq m$. Systems of the form (2.1.1) appear in many applications and have attracted a lot of research, see [1] for a comprehensive survey. If $\mathcal{A}$ in (2.1.1) is nonsingular, then the block 2×2 linear system (2.1.1) is called a generalized saddle point system. In most applications such as fluid dynamics (Stokes problem), incompressible elasticity, constrained optimization and so on, we have $A = A^T, C = B$ and $D = 0$; for some problems such as linearized Navier-Stokes equations discritized by stabilized finite elements, we have $C \neq B$ and $D \neq 0$ but $\|D\|$

[1]School of Mathematical Sciences and Laboratory of Mathematics for Nonlinear Sciences, Fudan University, Shanghai, People's Republic of China; e-mail: zcao@fudan.edu.cn

Owe Axelsson and János Karátson (Eds)

is small in many such problems. Finally, we note that if one uses a Chebyshev collocation approach to discretize the Stokes equations, then one has a saddle point system with $C \neq B$.

The iterative solution of systems of the form (2.1.1) can be achieved by various methods. In particular, Krylov subspace methods such as MINRES or GMRES are applicable in symmetric or nonsymmetric cases, respectively. In most situations it is necessary to use preconditioning for such iterative methods. The role of the preconditioner is to reduce the required number of iterations such that it does not increase the amount of computation required at each iteration.

In view of the convergence properties of Krylov subspace methods such as MINRES or GMRES, a sufficient condition for a good preconditioner $\mathcal{P}$ is that the preconditioned matrix $\mathcal{T} = \mathcal{P}^{-1}\mathcal{A}$ has a minimal polynomial of low degree. This means that $\mathcal{T} = \mathcal{P}^{-1}\mathcal{A}$ has only a few distinct eigenvalues and that its Jordan canonical form has Jordan blocks of only small dimension. Thus, intuitively, a clustered spectrum (away from zero) often yields rapid convergence.

2 Block Schur Complement Preconditioners for Generalized Saddle Point Systems

The construction of good preconditioners for saddle point systems exploits the block structure of the system (2.1.1).

When $D = 0$ and A is nonsingular, Murphy *et al.* [2] proposed an ideal block diagonal Schur complement preconditioner

$$\mathcal{D}_{+,0} = \begin{bmatrix} A & 0 \\ 0 & CA^{-1}B^T \end{bmatrix} \tag{2.2.1}$$

and two ideal block triangular Schur complement preconditioners (extended by Ipsen [3] to the general case where $D \neq 0$):

$$\mathcal{T}_{+,0} = \begin{bmatrix} A & 0 \\ C & D+CA^{-1}B^T \end{bmatrix} \quad \text{and} \quad \mathcal{T}_{-,0} = \begin{bmatrix} A & 0 \\ C & -(D+CA^{-1}B^T) \end{bmatrix}. \tag{2.2.2}$$

They showed that the preconditioned matrix

$$\mathcal{D}_{+,0}^{-1}\mathcal{A} = \begin{bmatrix} I & A^{-1}B^T \\ (CA^{-1}B^T)^{-1}C & 0 \end{bmatrix} \tag{2.2.3}$$

is diagonalizable and has only three distinct eigenvalues: $1, \frac{1\pm\sqrt{5}}{2}$. Thus the convergence of a minimum residual method such as GMRES is ensured in at most three iterations.

For the two preconditioners in (2.2.2), the corresponding preconditioned matrices are

$$\mathcal{T}_{+,0}^{-1}\mathcal{A} = \begin{bmatrix} I & A^{-1}B^T \\ 0 & -I \end{bmatrix} \quad \text{and} \quad \mathcal{T}_{-,0}^{-1}\mathcal{A} = \begin{bmatrix} I & A^{-1}B^T \\ 0 & I \end{bmatrix},$$

which have exactly two distinct eigenvalues ± 1 and one eigenvalue 1, respectively. Therefore the convergence of either of the two preconditioned GMRES methods is ensured in at most two iterations. We note that $\mathcal{T}_{+,0}^{-1}\mathcal{A}$ is diagonalizable but $\mathcal{T}_{-,0}^{-1}\mathcal{A}$ is not and the degree of any of its Jordan blocks is not greater than 2.

For the practical use of these block Schur complement preconditioners, it is necessary to find inexpensive approximations of A and of the Schur complement $D+CA^{-1}B^T$.

2.1   Block diagonal Schur complement preconditioners

Motivated by the favourable properties of the ideal block diagonal preconditioner $\mathcal{D}_{+,0}$ in (2.2.1), de Sturler and Liesen [4] developed a general framework to construct and analyze a class of block diagonal Schur complement preconditioners and constraint preconditioners which are derived from splittings of the (1,1) block A in (2.1.1). Let

$$A = G - E$$

be a splitting of A, where G is nonsingular. Then the block diagonal Schur complement preconditioner is

$$\mathcal{D}_+ = \begin{bmatrix} G & 0 \\ 0 & CG^{-1}B^T \end{bmatrix}. \qquad (2.2.4)$$

Let $G^{-1}A = I - G^{-1}E \equiv I - T$. If G is a good approximation of A, then $\|T\| \equiv \|I - G^{-1}A\|$ will be small. The preconditioned matrix equals

$$\mathcal{D}_+^{-1}\mathcal{A} = \begin{bmatrix} I - T & G^{-1}B^T \\ (CG^{-1}B^T)^{-1}C & 0 \end{bmatrix} \equiv \begin{bmatrix} I - T & N \\ M & 0 \end{bmatrix},$$

where $M \in \mathbb{R}^{m \times n}, N \in \mathbb{R}^{n \times m}, MN = I_m, (NM)^2 = NM$. We note that NM is a projection. The preconditioned matrix $\mathcal{D}_+^{-1}\mathcal{A}$ can be written in the following perturbed form:

$$\mathcal{D}_+^{-1}\mathcal{A} = \begin{bmatrix} I & N \\ M & 0 \end{bmatrix} - \begin{bmatrix} T & 0 \\ 0 & 0 \end{bmatrix}. \qquad (2.2.5)$$

It has been shown in [4] that the matrix $\begin{bmatrix} I & N \\ M & 0 \end{bmatrix}$ is diagonalizable and the corresponding eigenvalues and eigenvectors have been derived there.

From (2.2.5) and using the Bauer-Fike theorem [5], the following result was given.

Theorem 1.1. [4] *Let $U_1 \equiv [u_1, \ldots, u_{n-m}]$ be an orthonormal basis of $\mathcal{N}(NM)$ and $U_2 \equiv [u_{n-m+1}, \ldots, u_n]$ be an orthonomal basis of $\mathbb{R}(NM)$. For each eigenvalue λ of the preconditioner matrix $\mathcal{D}_+^{-1}\mathcal{A}$ there is an eigenvalue μ ($\mu = 1$ or $\frac{1 \pm \sqrt{5}}{2}$), such that*

$$|\lambda - \mu| \leq \sqrt{2} \left(\frac{1 + \omega_1}{1 - \omega_1} \right)^{\frac{1}{2}} \|T\|,$$

where ω_1 is the largest singular value of $U_1^T U_2$.

Cao [6] proposed an ideal block diagonal Schur complement preconditioner

$$\mathcal{D}_{-,0} = \begin{bmatrix} A & 0 \\ 0 & -CA^{-1}B^T \end{bmatrix} \qquad (2.2.6)$$

and showed that the preconditioned matrix $\mathcal{D}_{-,0}^{-1}\mathcal{A}$ is diagonalizable and has only three distinct eigenvalues: $1, \frac{1 \pm i\sqrt{3}}{2}$.

Motivated by the good properties of the ideal block diagonal preconditioner $\mathcal{D}_{-,0}$ in (2.2.6) and using the splitting $A = G - E$, Cao proposed the following block diagonal Schur complement preconditioner:

$$\mathcal{D}_- = \begin{bmatrix} G & 0 \\ 0 & -CG^{-1}B^T \end{bmatrix}. \qquad (2.2.7)$$

Here the preconditioned matrix is

$$\mathcal{D}_-^{-1}\mathcal{A} = \begin{bmatrix} I-T & N \\ -M & 0 \end{bmatrix} = \begin{bmatrix} I & N \\ -M & 0 \end{bmatrix} - \begin{bmatrix} T & 0 \\ 0 & 0 \end{bmatrix}.$$

The following result has been obtained.

Theorem 1.2 [6] *Let* $U_1 \equiv [u_1,\dots,u_{n-m}]$ *form an orthonormal basis of* $\mathcal{N}(NM)$ *and* $U_2 \equiv [u_{n-m+1},\dots,u_n]$ *form an orthonomal basis of* $\mathbb{R}(NM)$. *For each eigenvalue* λ *of the preconditioner matrix* $\mathcal{D}_-^{-1}\mathcal{A}$ *there is an eigenvalue* μ $(\mu = 1 \ or \ \frac{1\pm i\sqrt{3}}{2})$ *such that*

$$|\lambda - \mu| \le \sqrt{2}\left(\frac{1+\omega_1}{1-\omega_1}\right)^{\frac{1}{2}} \|T\|,$$

where ω_1 *is the largest singular value of* $U_1^T U_2$, *i.e.* $\omega_1 = \|U_1^T U_2\|$.

For the generalized saddle point matrix $\mathcal{A}$ in (2.1.1) (i.e., when $D \ne 0$), Siefert and de Sturler [7] studied the block diagonal Schur complement preconditioner (cf. (2.2.4))

$$\mathcal{D}_{+,D} = \begin{bmatrix} G & 0 \\ 0 & D+CG^{-1}B^T \end{bmatrix}$$

and extended Theorem 1.1 to the preconditioned matrix $\mathcal{D}_{+,D}^{-1}\mathcal{A}$.

2.2   Block triangular Schur complement preconditioners

Motivated by the two ideal block triangular Schur complement preconditioners

$$\mathcal{T}_{+,0} = \begin{bmatrix} A & 0 \\ C & D+CA^{-1}B^T \end{bmatrix} \quad \text{and} \quad \mathcal{T}_{-,0} = \begin{bmatrix} A & 0 \\ C & -(D+CA^{-1}B^T) \end{bmatrix}$$

and using the splitting $A = G - E$, Cao [8] proposed two block triangular Schur complement preconditioners:

$$\mathcal{T}_+ = \begin{bmatrix} G & 0 \\ C & D+CG^{-1}B^T \end{bmatrix} \quad \text{and} \quad \mathcal{T}_- = \begin{bmatrix} G & 0 \\ C & -(C+CG^{-1}B^T) \end{bmatrix}, \tag{2.2.8}$$

respectively.

Let $G^{-1}A = I - G^{-1}E \equiv I - T$. Denoting $S = D + CG^{-1}B^T$, we have the preconditioned matrices

$$\begin{aligned}
\mathcal{T}_\pm^{-1}\mathcal{A} &= \begin{bmatrix} G^{-1} & 0 \\ \mp S^{-1}CG^{-1} & \pm S^{-1} \end{bmatrix} \begin{bmatrix} A & B^T \\ C & -D \end{bmatrix} \\
&= \begin{bmatrix} G^{-1}A & G^{-1}B^T \\ \pm S^{-1}C(I-G^{-1}A) & \mp I_m \end{bmatrix} = \begin{bmatrix} I_n - T & G^{-1}B^T \\ \pm S^{-1}CT & \mp I_m \end{bmatrix}.
\end{aligned} \tag{2.2.9}$$

The preconditioned matrix $\mathcal{T}_+^{-1}\mathcal{A}$ can be written in the following perturbation form:

$$\mathcal{T}_+^{-1}\mathcal{A} = \begin{bmatrix} I_n & G^{-1}B^T \\ 0 & -I_m \end{bmatrix} - \begin{bmatrix} I_n & 0 \\ -S^{-1}C & 0 \end{bmatrix} \begin{bmatrix} T & 0 \\ 0 & 0 \end{bmatrix}. \tag{2.2.10}$$

It has been shown [8] that the matrix $\begin{bmatrix} I_n & G^{-1}B^T \\ 0 & -I_m \end{bmatrix}$ is diagonalizable and the eigenvector matrices corresponding to $\lambda = 1$ and $\lambda = -1$ are

$$\begin{bmatrix} I_n \\ 0 \end{bmatrix} \quad \text{and} \quad \begin{bmatrix} -\frac{1}{2}G^{-1}B^T \\ I_m \end{bmatrix},$$

respectively. From (2.2.10) and using the Bauer-Fike theorem [5] the following result was given.

Theorem 1.3 [8] *Let $S = D + CG^{-1}B^T$. Each eigenvalue λ of the block triangular Schur complement preconditioned matrix $\mathcal{T}_+^{-1}\mathcal{A}$ satisfies*

$$|\lambda \pm 1| \leq \left[1 + \frac{\|G^{-1}B^T\|}{4}\left(\frac{\|G^{-1}B^T\|}{2} + \sqrt{\frac{\|G^{-1}B^T\|^2}{4} + 4} \right) \right] \sqrt{1 + \|S^{-1}C\|^2}\|T\|$$

$$\leq \left(1 + \frac{1}{2}\|G^{-1}B^T\| \right)^2 \sqrt{1 + \|S^{-1}C\|}\|T\|,$$

where $|\lambda \pm 1|$ means $|\lambda + 1|$ or $|\lambda - 1|$.

The preconditioned matrix $\mathcal{T}_-^{-1}\mathcal{A}$ can be written in the following perturbation form:

$$\mathcal{T}_-^{-1}\mathcal{A} = \begin{bmatrix} I_n & G^{-1}B^T \\ 0 & I_m \end{bmatrix} - \begin{bmatrix} I_n & 0 \\ S^{-1}C & 0 \end{bmatrix}\begin{bmatrix} T & 0 \\ 0 & 0 \end{bmatrix}. \tag{2.2.11}$$

It has been shown in [8] that the matrix $\begin{bmatrix} I_n & G^{-1}B^T \\ 0 & I_m \end{bmatrix}$, which has only one eigenvalue 1, is not diagonalizable but the degree of its minimal polynomial is 2. Let $G^{-1}B^T = U\Sigma V^T$ be the singular value decomposition of $G^{-1}B^T$, where $\Sigma \in \mathbb{R}^{m \times n}, \Sigma^T = [\Sigma_m, 0]$ and $\Sigma_m = diag(\sigma_1, \ldots, \sigma_m)$ with $\sigma_j > 0, j = 1, \ldots, m$. It has been shown that the Jordan basis of the matrix $\begin{bmatrix} I_n & G^{-1}B^T \\ 0 & I_m \end{bmatrix}$ is $\begin{bmatrix} U & 0 \\ 0 & V \end{bmatrix}Q$, where

$$Q = \begin{bmatrix} e_1^{(n)} & 0 & \cdots & e_m^{(n)} & 0 & e_{m+1}^{(n)} & \cdots & e_n^{(n)} \\ 0 & \sigma_1^{-1}e_1^{(m)} & \cdots & 0 & \sigma_m^{-1}e_m^{(m)} & 0 & \cdots & 0 \end{bmatrix},$$

further, $e_j^{(n)} \in \mathbb{R}^n$ and $e_j^{(m)} \in \mathbb{R}^m$ are the jth unit vectors of n dimension and m dimension, respectively. Using a well-known result in matrix perturbation theory [5] and from (2.2.11), the following result was given:

Theorem 1.4 [8] *Let $S = D + CG^{-1}B^T$. Each eigenvalue λ of the block triangular preconditioned matrix $\mathcal{T}_-^{-1}\mathcal{A}$ satisfies*

$$|\lambda - 1| \leq \frac{1}{2}\left(\sqrt{4c_l + c_l^2\|T\|} + c_l\sqrt{\|T\|} \right)\sqrt{\|T\|},$$

where $c_l = \max\{1, \sigma_{min}^{-1}\}\max\{1, \sigma_{max}\}\sqrt{1 + \|S^{-1}C\|^2}$, further, σ_{min} and σ_{max} are the smallest and largest singular values of $G^{-1}B^T$, respectively.

2.3 Constraint Schur complement preconditioners

Let us consider a symmetric saddle point matrix

$$\mathcal{A} = \begin{bmatrix} A & B^T \\ B & 0 \end{bmatrix}.$$

Keller *et al.* [9] investigated the use of (so called constraint) preconditioners of the form

$$\mathcal{G}_0 = \begin{bmatrix} G & B^T \\ B & 0 \end{bmatrix}$$

with a CG-like method to solve the symmetric saddle point system, where $\mathcal{G}_0$ is also assumed to be symmetric. They determined the eigensolution distribution of the preconditioned matrix $\mathcal{G}_0^{-1}\mathcal{A}$ and an upper bound of the degree of its minimal polynomial. Cao [10] extended the constraint preconditioner to nonsymmetric saddle point matrices, i.e. when A and G may be nonsymmetric, while Dollar [11] extended the symmetric constraint preconditioner to symmetric saddle point matrices with the (2,2) block being symmetric negative semidefinite.

For the generalized saddle point matrix $\mathcal{A}$ in (2.1.1), Cao [12,13] studied constraint Schur complement preconditioners of the form

$$\mathcal{G} = \begin{bmatrix} G & B^T \\ C & -D \end{bmatrix}, \tag{2.2.12}$$

where G is derived from a splitting of A. Results concerning the eigensolution distribution of the preconditioned matrix $\mathcal{G}^{-1}\mathcal{A}$ and its minimal polynomial were given. Assume that the saddle point matrix $\mathcal{A}$ in (2.1.1) satisfies the following:

B and C are of full rank,

$rank(D) = p, \ 0 \le p \le m.$

Then, corresponding to $p = m, p = 0$ and $0 < p < m$, the following three results were given.

Theorem 1.5 [13] *Let $\mathcal{A}$ and $\mathcal{G}$ be two nonsingular generalized saddle point matrices of the form given in (2.1.1) and (2.2.12), respectively. Assume that $B \in \mathbb{R}^{m \times n}$ and $C \in \mathbb{R}^{m \times n}$ are of full rank, and D is nonsingular. Then the constraint preconditioned matrix $\mathcal{G}^{-1}\mathcal{A}$ has*
1. m eigenvalues of unit value;
2. n eigenvalues that are defined by the generalized eigenvalue problem

$$(A + B^T D^{-1} C)x = \lambda(G + B^T D^{-1} C)x.$$

Additionally, the matrix $\mathcal{G}^{-1}\mathcal{A}$ has $m + i + j$ linearly independent eigenvectors:
1. m eigenvectors of the form $[0^T, y^T]^T$ corresponding to the eigenvalue one of $\mathcal{G}^{-1}\mathcal{A}$;
2. $i \ (0 \le i \le n)$ eigenvectors of the form $[x^T, y^T]^T$ corresponding to the eigenvalue one of $\mathcal{G}^{-1}\mathcal{A}$, where the components x arise from the generalized eigenvalue problem $Ax = Gx$;
3. $j \ (0 \le j \le n)$ eigenvectors of the form $[x^T, y^T]^T$ corresponding to the eigenvalues of $\mathcal{G}^{-1}\mathcal{A}$ not equal to one, where the components x arise from the generalized eigenvalue problem $(A + B^T D^{-1} C)x = \lambda(G + B^T D^{-1} C)x$ with $\lambda \ne 1$.

Theorem 1.6 [13] *Let $\mathcal{A}$ and $\mathcal{G}$ be two nonsingular generalized saddle point matrices of the form given in (2.1.1) and (2.2.12), respectively. Assume that $B \in \mathbb{R}^{m \times n}$ and $C \in \mathbb{R}^{m \times n}$ are of full rank, and $D = 0$. Let the columns of $V_2 \in \mathbb{R}^{n \times (n-m)}$ and $Z_2 \in \mathbb{R}^{n \times (n-m)}$ span the nullspaces of B and C, respectively. Then the constraint preconditioned matrix $\mathcal{G}^{-1}\mathcal{A}$ has*

1. 2m eigenvalues of unit value;
2. $n-m$ eigenvalues that are defined by the generalized eigenvalue problem

$$V_2^T A Z_2 \widehat{x}_2 = \lambda V_2^T G Z_2 \widehat{x}_2.$$

Additionally, the matrix $G^{-1}\mathcal{A}$ has $m+i+j$ linearly independent eigenvectors:
1. m eigenvectors of the form $[0^T, y^T]^T$ corresponding to the eigenvalue one of $G^{-1}\mathcal{A}$;
2. i ($0 \leq i \leq n$) eigenvectors of the form $[x^T, y^T]^T$ corresponding to the eigenvalue one of $G^{-1}\mathcal{A}$, where the components x arise from the generalized eigenvalue problem $Ax = Gx$;
3. j ($0 \leq j \leq n-m$) eigenvectors of the form $[\widehat{x}_2^T Z_2^T, y^T]^T$ corresponding to the eigenvalues of $G^{-1}\mathcal{A}$ not equal to one, where the components $\widehat{x}_2$ arise from the generalized eigenvalue problem $V_2^T A Z_2 \widehat{x}_2 = V_2^T G Z_2 \widehat{x}_2$ with $\lambda \neq 1$.

Theorem 1.7 [13] *Let $\mathcal{A}$ and G be two nonsingular generalized saddle point matrices of the form given in (2.1.1) and (2.2.12), respectively. Assume that $B \in \mathbb{R}^{m \times n}$ and $C \in \mathbb{R}^{m \times n}$ are of full rank, and D has rank of p, where $0 < p < m$. Let*

$$D = Q\Delta P^T = [Q_1, Q_2] \begin{bmatrix} \Delta_p & 0 \\ 0 & 0 \end{bmatrix} \begin{bmatrix} P_1^T \\ P_2^T \end{bmatrix} \qquad (2.2.13)$$

be a singular value decomposition of D, where $Q \equiv [Q_1, Q_2]$ and $P \equiv [P_1, P_2]$ are orthogonal matrices of order m, $Q_1, P_1 \in \mathbb{R}^{m \times p}$ and $Q_2, P_2 \in \mathbb{R}^{m \times (m-p)}$, while Δ_p is a $p \times p$ diagonal matrix with positive diagonal entries. Let the columns of $M_2 \in \mathbb{R}^{n \times (n-m+p)}$ and $N_2 \in \mathbb{R}^{n \times (n-m+p)}$ span the nullspaces of $P_2^T B$ and $Q_2^T C$, respectively. Then the constraint preconditioned matrix $G^{-1}\mathcal{A}$ has
1. $2m - p$ eigenvalues of unit value;
2. $n - m + p$ eigenvalues that are defined by the generalized eigenvalue problem

$$M_2^T (A + B^T P_1 \Delta_p^{-1} Q_1^T C) N_2 \widehat{x}_2 = \lambda M_2^T (G + B^T P_1 \Delta_p^{-1} Q_1^T C) N_2 \widehat{x}_2. \qquad (2.2.14)$$

Additionally, the matrix $G^{-1}\mathcal{A}$ has $m+i+j$ linearly independent eigenvectors:
1. m eigenvectors of the form $[0^T, y^T]^T$ corresponding to the eigenvalue one of $G^{-1}\mathcal{A}$;
2. i ($0 \leq i \leq n$) eigenvectors of the form $[x^T, y^T]^T$ corresponding to the eigenvalue one of $G^{-1}\mathcal{A}$, where the components x arise from the generalized eigenvalue problem $Ax = Gx$;
3. j ($0 \leq j \leq n-m+p$) eigenvectors of the form $[\widehat{x}_2^T N_2^T, y^T]^T$ corresponding to the eigenvalues of $G^{-1}\mathcal{A}$ not equal to one, where the components $\widehat{x}_2$ arise from the generalized eigenvalue problem (cf. (2.2.14))

$$M_2^T (A + B^T P_1 \Delta_p^{-1} Q_1^T C) N_2 \widehat{x}_2 = \lambda M_2^T (G + B^T P_1 \Delta_p^{-1} Q_1^T C) N_2 \widehat{x}_2$$

with $\lambda \neq 1$.

Concerning the minimal polynomial of the preconditioned matrix $G^{-1}\mathcal{A}$, the following result was given.

Theorem 1.8 [13] *For $0 < p < m$ let $S_3 = [M_2^T (G + B^T P_1 \Delta_p^{-1} Q_1^T C) N_2]^{-1} [M_2^T (A + B^T P_1 \Delta_p^{-1} Q_1^T C) N_2]$ (where P_1, Q_1, Δ_p, M_2 and N_2 cf. Theorem 1.7); for $p = 0$ let $S_2 = (V_2^T G Z_2)^{-1} (V_2^T A Z_2)$, where V_2 and Z_2 are orthonormal bases of $\mathcal{N}(B)$ and $\mathcal{N}(C)$, respectively, and for $p = m$ let $S_1 = (G + B^T D^{-1} C)^{-1} (A + B^T D^{-1} C)$. If the degrees of the minimal polynomials of S_i are k_i, $i = 1, 2, 3$, respectively, then the degree of the minimal polynomial of the preconditioned matrix $G^{-1}\mathcal{A}$ is at most*

$$\begin{aligned}
k_1 + 1 &\leq n + 1, & \text{when} \quad p &= m; \\
k_2 + 2 &\leq n - m + 2, & \text{when} \quad p &= 0; \\
k_3 + 3 &\leq n - m + p + 3, & \text{when} \quad 0 &< p < m.
\end{aligned}$$

For $D = 0$ the saddle point matrix $\mathcal{A}$ and the constraint preconditioner are

$$\mathcal{A} = \begin{bmatrix} A & B^T \\ C & 0 \end{bmatrix} \quad \text{and} \quad \begin{bmatrix} G & B^T \\ C & 0 \end{bmatrix}, \tag{2.2.15}$$

respectively. Let $A = G - E$ and denote $G^{-1}E = T$. If G is a good approximation of A, then $\|T\| = \|I - G^{-1}A\|$ is small. The following perturbation result was given by de Sturler and Liesen [4].

Theorem 1.9 *Let $M = (CG^{-1}B^T)^{-1}C, N = G^{-1}B^T$ and $U_1 = [u_1, \ldots, u_{n-m}]$ form an orthonormal basis of $\mathcal{N}(NM)$, $U_2 = [n_{n-m+1}, \ldots, u_n]$ form an orthonormal basis of $\mathbb{R}(NM)$. Then each eigenvalues λ of the constraint preconditioned matrix $G^{-1}\mathcal{A}$ satisfies*

$$|\lambda - 1| \leq \frac{\|T\|}{\sqrt{1 - \omega_1^2}},$$

where ω_1 is the largest singular value of $U_1^T U_2$, i.e. $\|U_1^T U_2\|$.

Siefer and de Sturler [7] and Cao [12] have extended this result to the general case where the (2,2) block in $\mathcal{A}$ and $\mathcal{G}$ in (2.2.15) may be nonzero.

2.4 Implementations

For the implementation of these preconditioners, we choose G as an incomplete LU factorization of A

$$A = LU + R, \quad G = LU$$

with a drop tolerance τ [14] which controls the quality of the approximation G to the (1,1) block A of $\mathcal{A}$.

For the block diagonal Schur complement preconditioner $\mathcal{D}_{\pm}$, the vector $z = \mathcal{D}_{\pm}^{-1}v$, or, equivalently, the solution $z = [z_1^T, z_2^T]^T$ of the linear system

$$\begin{bmatrix} G & 0 \\ 0 & \pm(D + CG^{-1}B^T) \end{bmatrix} \begin{bmatrix} z_1 \\ z_2 \end{bmatrix} = \begin{bmatrix} v_1 \\ v_2 \end{bmatrix}$$

can be obtained by the following algorithm.

Algorithm $\mathcal{D}_{\pm}$. For a given vector $v = [v_1^T, v_2^T]^T$ compute the vector $z = [z_1^T, z_2^T]^T = \mathcal{D}_{\pm}^{-1}v$
(1) $z_1 = U^{-1}(L^{-1}v_1)$;
(2) Solve $(D + CU^{-1}L^{-1}B^T)z_2 = \pm v_2$.

For the block triangular Schur complement preconditioner $\mathcal{T}_{\pm}$, the vector $z = \mathcal{T}_{\pm}^{-1}v$, or, equivalently, the solution $z = [z_1^T, z_2^T]^T$ of the linear system

$$\begin{bmatrix} G & 0 \\ C & \pm(D + CG^{-1}B^T) \end{bmatrix} \begin{bmatrix} z_1 \\ z_2 \end{bmatrix} = \begin{bmatrix} v_1 \\ v_2 \end{bmatrix}$$

can be obtained by the following algorithm.

Algorithm $\mathcal{T}_{\pm}$. For a given vector $v = [v_1^T, v_2^T]^T$ compute the vector $z = [z_1^T, z_2^T]^T = \mathcal{T}_{\pm}^{-1}v$

(1) $z_1 = U^{-1}(L^{-1}v_1)$;

(2) Solve $(D + CU^{-1}L^{-1}B^T)z_2 = \pm(v_2 - Cz_1)$.

For the constraint preconditioner $\mathcal{G}$ we note that

$$\mathcal{G} \equiv \begin{bmatrix} G & B^T \\ C & -D \end{bmatrix} = \begin{bmatrix} G & 0 \\ C & -(D + CG^{-1}B^T) \end{bmatrix} \begin{bmatrix} I & G^{-1}B^T \\ & I \end{bmatrix},$$

the vector $z = \mathcal{G}^{-1}v$ or, equivalently, the solution $z = [z_1^T, z_2^T]^T$ of the linear system $\mathcal{G}\begin{bmatrix} z_1 \\ z_2 \end{bmatrix} = \begin{bmatrix} v_1 \\ v_2 \end{bmatrix}$ can be obtained by the following algorithm.

Algorithm $\mathcal{G}$. For a given vector $v = [v_1^T, v_2^T]^T$ compute the vector $z = [z_1^T, z_2^T]^T = \mathcal{G}^{-1}v$

(1) $z_1 = U^{-1}(L^{-1}v_1)$;

(2) $t_2 = Cz_1 - v_2$;

(3) Solve $(D + CU^{-1}L^{-1}B^T)z_2 = t_2$;

(4) $t_1 = U^{-1}(L^{-1}(B^Tz_2))$;

(5) $z_1 = z_1 - t_1$.

These algorithms show that the crucial step in order to achieve fast convergence of the preconditioned GMRES is the efficient solution of the Schur complement system

$$(D + CU^{-1}L^{-1}B^T)y = w. \tag{2.2.16}$$

In order to solve the Schur complement system efficiently, we consider the special case of the matrix $\mathcal{A}$ in (2.1.1) with $C = B$ and $D = 0$, i.e., we consider saddle point systems of the form

$$\begin{bmatrix} A & B^T \\ B & 0 \end{bmatrix} \begin{bmatrix} x \\ y \end{bmatrix} = \begin{bmatrix} f \\ g \end{bmatrix}. \tag{2.2.17}$$

Such systems arise, for example, in the MAC discretization [15] or in the mixed finite element discretization by stable element pairs of the Oseen equations [16]. In this case, the Schur complement system (3.1) is simplified as

$$(BG^{-1}B^T)y = w. \tag{2.2.18}$$

For the efficient approximate solution of the Schur complement system (2.2.18), Elman *et al.* [17] used a strategy based on approximate commutators, i.e., they aimed at finding an approximate solution $\widehat{G}$ of X to the matrix equation

$$B^TX = GB^T. \tag{2.2.19}$$

The least squares solution of (2.2.19) is

$$\widehat{G} = (BB^T)^{-1}BGB^T.$$

Then, the approximate solution $\widehat{y}$ of the Schur complement system (2.2.18) is

$$\widehat{y} = (BB^T)^{-1}(BGB^T)(BB^T)^{-1}w. \tag{2.2.20}$$

Let $BB^T = L_bU_b$ be the *LU* factorization of the matrix $BB^T \in \mathbb{R}^{m \times m}$, then we have the following algorithm:

Algorithm SC. For a given vector w compute the vector $\widehat{y} = (L_bU_b)^{-1}(BLUB^T)(L_bU_b)^{-1}w$ as an approximate solution of (2.2.18).

(1) $t_1 = U_b^{-1}(L_b^{-1}w)$;
(2) $t_2 = L(U(B^T t_1))$;
(3) $\hat{y} = U_b^{-1}(L_b^{-1}(Bt_2))$.

Let the (2,2) block of $\mathcal{A}$ in (2.1.1) satisfy $-D \neq 0$ such that $\|D\|$ is small. The corresponding saddle point system

$$\begin{bmatrix} A & B^T \\ B & -D \end{bmatrix} \begin{bmatrix} x \\ y \end{bmatrix} = \begin{bmatrix} f \\ g \end{bmatrix}$$

arises, for example, in the mixed finite element discretization of the Oseen equations by stabilized element pairs [16,18,19]. In this case (cf. (2.2.16)), the Schur complement system is

$$(D + BU^{-1}L^{-1}B^T)y = w, \tag{2.2.21}$$

with $D \neq 0, \|D\|$ being small. We can also use $\hat{y}$ in (2.2.20) (with $G = LU$) as the approximate solution of (2.2.21).

If we use Algorithm SC to solve Schur complement system (2.2.18) or (2.2.21) approximately, then we obtain the following algorithms. They provide the implementation of the preconditioners $\mathcal{D}_\pm, \mathcal{T}_\pm$ and $\mathcal{G}$, which are used in the preconditioned GMRES to solve saddle point systems of the form

$$\begin{bmatrix} A & B^T \\ B & -D \end{bmatrix} \begin{bmatrix} x \\ y \end{bmatrix} = \begin{bmatrix} f \\ g \end{bmatrix}, \tag{2.2.22}$$

where $D = 0$, or $D \neq 0$ but $\|D\|$ is small.

Algorithm $\mathcal{D}_\pm ac$. For a given vector $v = [v_1^T, v_2^T]^T$ compute the vector $z = [z_1^T, z_2^T]^T = \mathcal{D}_\pm^{-1}v$.
(1) $z_1 = U^{-1}(L^{-1}v_1)$;
(2) $t_1 = U_b^{-1}(L_b^{-1}v_2)$;
(3) $t_2 = L(U(B^T t_1))$;
(4) $z_2 = \pm U_b^{-1}(L_b^{-1}(Bt_2))$.

Algorithm $\mathcal{T}_\pm ac$. For a given vector $v = [v_1^T, v_2^T]^T$ compute the vector $z = [z_1^T, z_2^T]^T$.
$= \mathcal{T}_\pm^{-1}v$
(1) $t_1 = U_b^{-1}(L_b^{-1}v_2)$;
(2) $t_2 = L(U(B^T t_1))$;
(3) $z_2 = \pm U_b^{-1}(L_b^{-1}(Bt_2))$;
(4) $z_1 = U^{-1}(L^{-1}(v_1 - B^T z_2))$.

Algorithm $\mathcal{G}ac$. For a given vector $v = [v_1^T, v_2^T]^T$ compute the vector $z = [z_1^T, z_2^T]^T = \mathcal{G}^{-1}v$.
(1) $z_1 = U^{-1}(L^{-1}v_1)$;
(2) $t_2 = Bz_1 - v_2$;
(3) $t_1 = U_b^{-1}(L_b^{-1}t_2))$;
(4) $t_3 = L(U(B^T t_1))$;
(5) $z_2 = U_b^{-1}(L_b^{-1}(Bt_3))$;
(6) $t_1 = U^{-1}(L^{-1}(B^T z_2))$;
(6) $z_1 = z_1 - t_1$.

We give a numerical example to compare the performance of the five preconditioners that we have discussed. The problem under consideration arises from the linearization of the steady-state Navier-Stokes equations, i.e., the Oseen problems of the following form:

$$-v\Delta u + w \cdot \nabla u + grad \ p = f \qquad \text{in } \Omega \qquad (2.2.23)$$
$$div \ u = 0$$

with suitable boundary conditions on $\partial\Omega$, where $\Omega \subset \mathbb{R}^2$ is a bounded domain and w is a given divergence free field. The scalar v is the viscosity, the vector field u represents the velocity, and p denotes the pressure.

The test problem is a "leaky" two-dimensional lid-driven cavity problem in a square domain ($0 \le x \le 1 : 0 \le y \le 1$). The boundary conditions are $u_x = u_y = 0$ on the three fixed walls ($x = 0, y = 0, x = 1$), and $u_x = 1, u_y = 0$ on the moving wall ($y = 1$). On the constructing coefficient matrix $\mathcal{A}$, we use the circulating wind: $w_x = 8x(x-1)(1-2y), w_y = 8(2x-1)y(y-1)$ (cf. [15]).

To discretize (2.2.23), we use two methods. One is a finite element subdivision based on $ne \times ne$ uniform grids of square elements, the mixed finite element used is the bilinear-constant velocity-pressure ($Q_1 - P_0$) element [18] with global stabilization or local stabilization [18,19]. The second is a "marker and cell" (MAC) finite difference scheme [15] based on an $ne \times ne$ uniform grids of square meshes. We note that in these two methods of discretization $C = B$ in the matrix $\mathcal{A}$ in (2.1.1), furthermore, $D = 0$ if the MAC scheme is used and if $D \ne 0$, $\|D\|$ is small.

Figures 1–5 plot the eigenvalues of the five preconditioned matrices arising from MAC scheme for drop tolerance $\tau = 0.01, v = 1$ and $h(= \frac{1}{ne}) = \frac{1}{16}$ (thus, the number of the eigenvalues is 736). Among them Fig.1 plots the eigenvalues of the block diagonal preconditioned matrix $\mathcal{D}_+^{-1}\mathcal{A}$. Fig.2–5 plot those of $\mathcal{D}_-^{-1}\mathcal{A}, \mathcal{T}_+^{-1}\mathcal{A}, \mathcal{T}_-^{-1}\mathcal{A}$ and $\mathcal{G}^{-1}\mathcal{A}$, respectively.

Fig.(1). Eigenvalues of block diagonal preconditioned matrix $\mathcal{D}_+^{-1}\mathcal{A}$

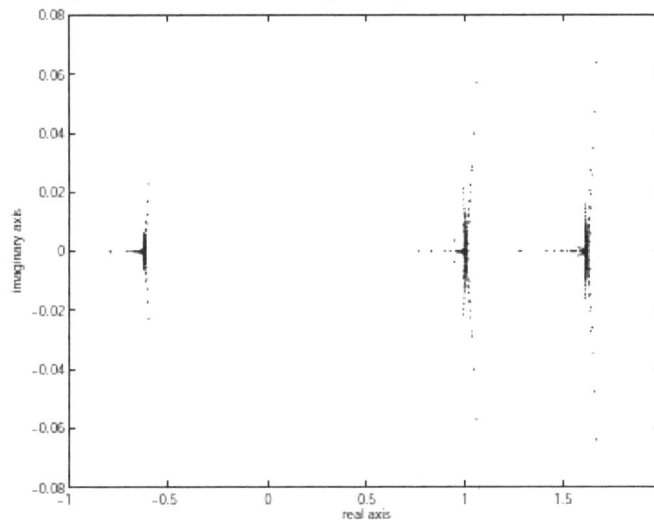

Fig.(2). Eigenvalues of block diagonal preconditioned matrix $\mathcal{D}_-^{-1}\mathcal{A}$

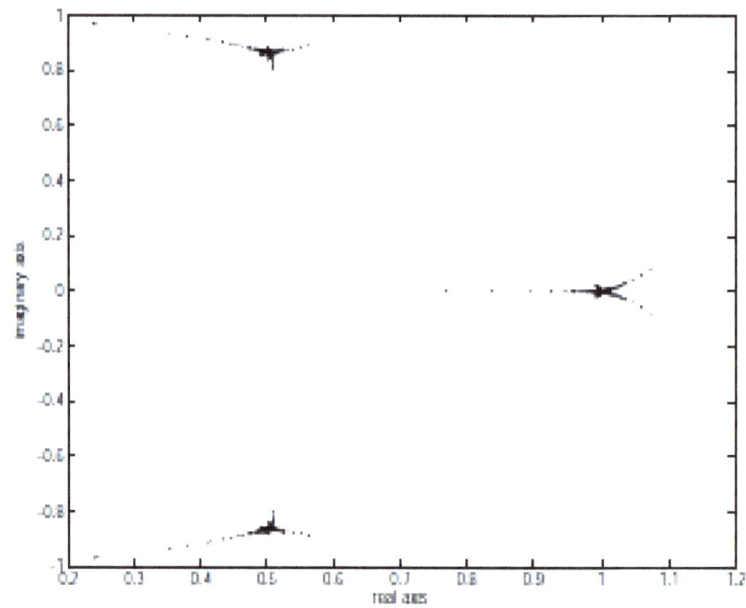

Fig.(3). Eigenvalues of block triangular preconditioned matrix $\mathcal{T}_+^{-1}\mathcal{A}$

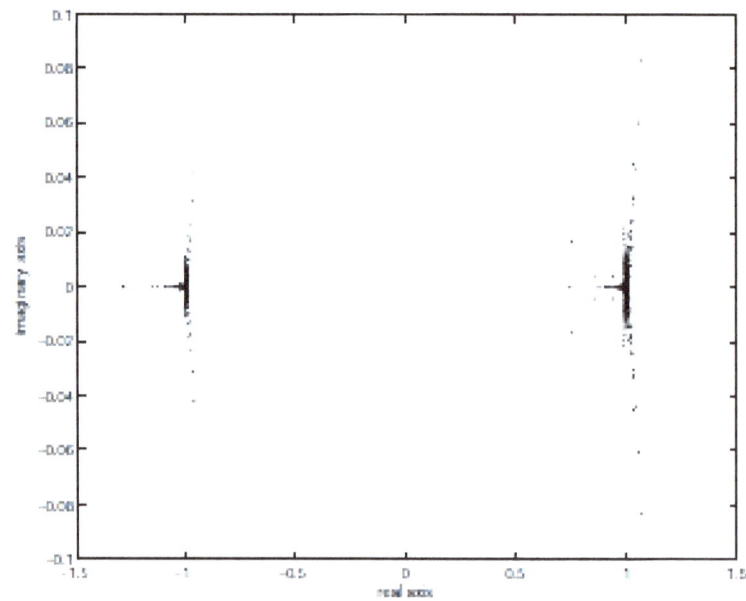

Fig.(4). Eigenvalues of block triangular preconditioned matrix $\mathcal{T}_-^{-1}\mathcal{A}$

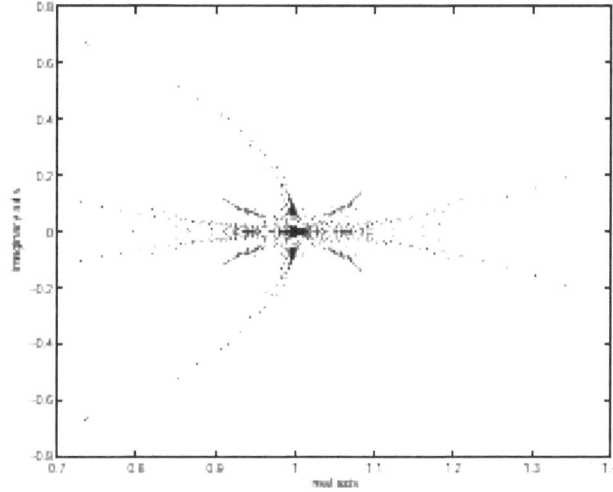

Fig.(5). Eigenvalues of constraint preconditioned matrix $\mathcal{G}^{-1}\mathcal{A}$

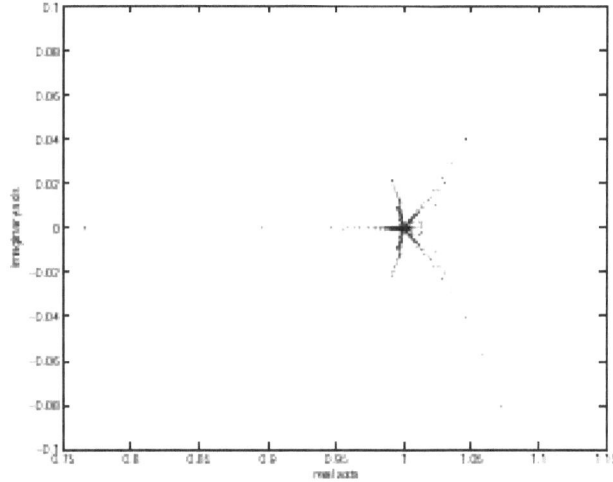

• From Fig.1 we can see that the eigenvalues of $\mathcal{D}_+^{-1}\mathcal{A}$ cluster around 1 or $\frac{1\pm\sqrt{5}}{2}$ (cf. Theorem 1.1).

• From Fig.2 we can see that the eigenvalues of $\mathcal{D}_-^{-1}\mathcal{A}$ cluster around 1 or $\frac{1\pm i\sqrt{3}}{2}$ (cf. Theorem 1.2). Moreover, the real part of each eigenvalue of $\mathcal{D}_-^{-1}\mathcal{A}$ is positive, i.e., $\mathcal{D}_-^{-1}\mathcal{A}$ is a positive stable matrix.

• From Fig.3 we can see that the eigenvalues of $\mathcal{T}_+^{-1}\mathcal{A}$ cluster around 1 or -1 (cf. Theorem 1.3).

• From Fig.4 we can see that the eigenvalues of $\mathcal{T}_-^{-1}\mathcal{A}$ cluster around 1 (cf. Theorem 1.4). Moreover, the real part of each eigenvalue of $\mathcal{T}_-^{-1}\mathcal{A}$ is positive, i.e., $\mathcal{T}_-^{-1}\mathcal{A}$ is a positive stable matrix.

• From Fig.5 we can see that the eigenvalues of $\mathcal{G}^{-1}\mathcal{A}$ cluster around 1. Moreover, comparing Fig.5 with Fig.4 we can see that the eigenvalues of $\mathcal{G}^{-1}\mathcal{A}$ cluster around 1 more closely than those of $\mathcal{T}_-^{-1}\mathcal{A}$.

We take $\nu = 0.01$ and $\tau = 0.001$ for $h = \frac{1}{128}$ to compare the five preconditioners $\mathcal{G}ac$, $\mathcal{T}_-ac$, $\mathcal{T}_+ac$, $\mathcal{D}_-ac$ and $\mathcal{D}_+ac$ for the MAC scheme; the local and global stabilizations. Fig. 6 shows the convergence history for the MAC scheme; Fig.7 shows the convergence history for the local stabilization and Fig.8 shows the convergence history for the global stabilization.

Fig.(6). Convergence history for MAC scheme. $\tau = 0.001, h = \frac{1}{128}, \nu = 0.01$

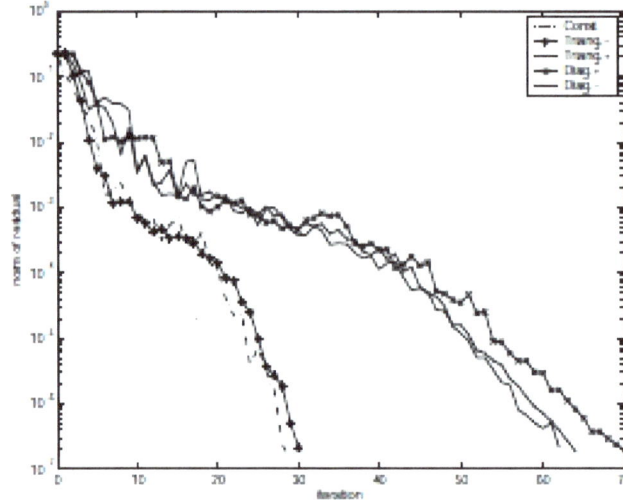

Fig.(7). Convergence history for local stabilization. $\tau = 0.001, h = \frac{1}{128}, \nu = 0.01$

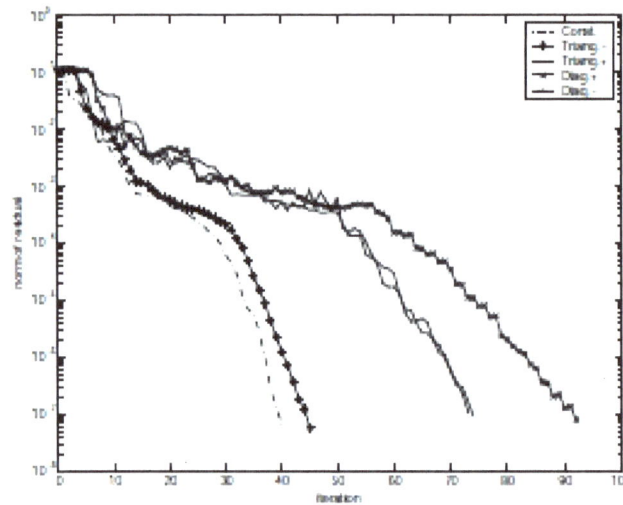

Fig. (8). Convergence history for global stabilization, $\tau = 0.001, h = \frac{1}{128}, v = 0.01$

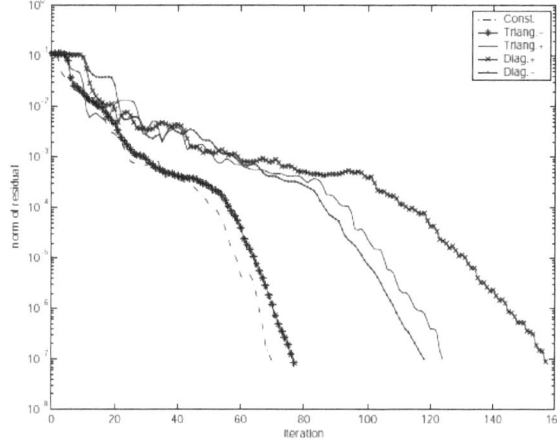

From Fig.6, Fig.7 and Fig.8 we can see that the performance of the preconditioner $\mathcal{G}ac$ or $\mathcal{T}\_ac$ is significantly more effective than that of each of the preconditioners $\mathcal{T}_+ac$, $\mathcal{D}\_ac$ and $\mathcal{D}_+ac$.

3   Augmentation Block Preconditioners for Saddle Point Systems with Singular (1,1) Blocks

Consider the saddle point system

$$\mathcal{A}\begin{bmatrix} x \\ y \end{bmatrix} \equiv \begin{bmatrix} A & B^T \\ C & 0 \end{bmatrix} \begin{bmatrix} x \\ y \end{bmatrix} = \begin{bmatrix} f \\ g \end{bmatrix}, \tag{2.3.1}$$

where $A \in \mathbb{R}^{n \times n}, B, C \in \mathbb{R}^{m,n}, m \leq n$. The matrix $\mathcal{A}$ is assumed to be nonsingular, whereas the matrix A is singular with high nullity. Since A is singular, it cannot be inverted and the Schur complement does not exist. One possible way of dealing with the system is by augmentation, i.e. by replacing A with $A + B^T W C$, where $W \in \mathbb{R}^{m \times m}$ such that $A + B^T W C$ is invertible. Then (2.3.1) is replaced by

$$\begin{bmatrix} A + B^T W C & B^T \\ C & 0 \end{bmatrix} \begin{bmatrix} x \\ y \end{bmatrix} = \begin{bmatrix} f + B^T W g \\ g \end{bmatrix}. \tag{2.3.2}$$

For solving this system we can use block Schur complement preconditioners discussed in Section 1. Another possible way to solve the system (2.3.1) is to use an augmentation block Schur complement-free preconditioner directly, for instance,

$$\mathcal{D}_{Aug} = \begin{bmatrix} A + B^T W^{-1} C & 0 \\ 0 & W \end{bmatrix},$$

where $W \in \mathbb{R}^{m \times m}$ is nonsingular and such that $A + B^T W^{-1} C$ is invertible.

3.1   Spectral analysis of augmentation block Schur complement preconditioners

In this section we consider the symmetric saddle point system

$$\mathcal{A}\begin{bmatrix} x \\ y \end{bmatrix} \equiv \begin{bmatrix} A & B^T \\ B & 0 \end{bmatrix} \begin{bmatrix} x \\ y \end{bmatrix} = \begin{bmatrix} f \\ g \end{bmatrix}, \tag{2.3.3}$$

where A is symmetric positive semidefinite and $\mathcal{A}$ is nonsingular, which implies that B has full rank. For solving the system (2.3.3) Golub *et al.* [20] presented a block diagonal symmetric positive definite Schur complement preconditioner based on augmentation:

$$\mathcal{D}(W) = \begin{bmatrix} A + B^T W B & 0 \\ 0 & B(A + B^T W B)^{-1} B^T \end{bmatrix}, \qquad (2.3.4)$$

where $W \in \mathbb{R}^{m \times m}$ is symmetric positive semidefinite and such that $A + B^T W B$ is symmetric positive definite. Algebraic analysis of this preconditioner is performed. The main conclusions are as follows.

• The eigenvalues of the preconditioned matrix $\mathcal{D}(W)^{-1} \mathcal{A}$ are clustered in two fixed intervals:

$$\left[-1, \frac{1 - \sqrt{5}}{2} \right] \cup \left[1, \frac{1 + \sqrt{5}}{2} \right],$$

whose ends are far from the origin.

• The eigenvalue 1 is of geometric multiplicity at least $n - m$ (i.e. there are at least $n - m$ linearly independent eigenvectors corresponding to the eigenvalue 1) and the multiplicity grows with the nullity of A.

Cao [21] has presented three augmentation block triangular Schur complement preconditioners:

$$\mathcal{T}_+(W) = \begin{bmatrix} A + B^T W B & B^T \\ & B(A + B^T W B)^{-1} B^T \end{bmatrix},$$

$$\mathcal{T}_-(W) = \begin{bmatrix} A + B^T W B & B^T \\ & -B(A + B^T W B)^{-1} B^T \end{bmatrix}$$

and

$$\mathcal{T}_{2-}(W) = \begin{bmatrix} A + B^T W B & 2B^T \\ & -B(A + B^T W B)^{-1} B^T \end{bmatrix}.$$

The following three results were given concerning their eigenvalue distribution, respectively.

Theorem 2.1 [21] *The eigenvalues of the preconditioned matrix $\mathcal{T}_+(W)^{-1} \mathcal{A}$ are all real and bounded within the two intervals*

$$\left[-\frac{\sqrt{5} + 1}{2}, -1 \right) \cup \left[\frac{\sqrt{5} - 1}{2}, 1 \right].$$

The eigenvalue 1 is of geometric multiplicity $n - m$, the corresponding eigenvectors are $[x_i^T, 0^T]^T$, where x_i, $i = 1, \ldots, n - m$, span the null space $\mathcal{N}(B)$. The eigenvalues $\frac{\sqrt{5} - 1}{2}$ and $-\frac{\sqrt{5} + 1}{2}$ are both of geometric multiplicity $s \equiv \text{nullity}(A)$, the corresponding eigenvectors are $[x_i^T, \frac{\sqrt{5} + 1}{2} x_i^T B^T (BM(W)^{-1} B^T)^{-1}]^T$, $i = 1, \ldots, s$ and $[x_i^T, -\frac{\sqrt{5} - 1}{2} x_i^T B^T (BM(W)^{-1} B^T)^{-1}]^T$, $i = 1, \ldots, s$, respectively, where x_i, $i = 1, \ldots, s$, span the null space $\mathcal{N}(A)$.

Theorem 2.2 [21] *The eigenvalues of $\mathcal{T}_-(W)^{-1} \mathcal{A}$ are bounded within the region*

$$\frac{1}{2} \leq Re(\lambda) \leq 1; \quad -\frac{\sqrt{3}}{2} \leq Im(\lambda) \leq \frac{\sqrt{3}}{2}.$$

The eigenvalue 1 is of geometric multiplicity $n - m$, the corresponding eigenvectors are $[x_i^T, 0^T]^T$, where x_i, $i = 1, \ldots, n - m$, span $\mathcal{N}(B)$. The eigenvalues $\frac{1 + i\sqrt{3}}{2}$ and $\frac{1 - i\sqrt{3}}{2}$ are both of geometric *multiplicity $s \equiv nullity(A)$, the corresponding eigenvectors are $[x_i^T, \frac{-1+i\sqrt{3}}{2} x_i^T B^T (BM(W)^{-1}B^T)^{-1}]^T$, $i = 1, \ldots, s$ and $[x_i^T, \frac{-1-i\sqrt{3}}{2} x_i^T B^T (BM(W)^{-1}B^T)^{-1}]^T$, $i = 1, \ldots, s$, respectively, where x_i, $i = 1, \ldots, s$, span the null space $\mathcal{N}(A)$.*

Theorem 2.3 [21] *The eigenvalues of the preconditioned matrix $\mathcal{T}_{2-}(W)^{-1}\mathcal{A}$ are all real and bounded within the interval*

$$\left(\frac{3 - \sqrt{5}}{2}, \frac{3 + \sqrt{5}}{2} \right).$$

The eigenvalue 1 is of multiplicity $n - m + 2s$, where $s = nullity(A)$, while the geometric multiplicity is $n - m + s$. The corresponding eigenvectors are $[x_i^T, 0^T]^T$, where x_i, $i = 1, \ldots, n - m$, span the null space $\mathcal{N}(B)$ and

$$[x_i^T, -x_i^T B^T (BM(W)^{-1}B^T)^{-1}]^T,$$

where x_i, $i = 1, \ldots, s$, span the null space $\mathcal{N}(A)$.

3.2   Augmentation block Schur complement-free preconditioners

In this section we consider the saddle point systems of the form (2.3.1). When $\mathcal{A}$ in (2.3.1) is symmetric, Greif *et al.* [22] have introduced a block diagonal Schur complement-free preconditioner based on augmentation:

$$\mathcal{M}_W = \begin{bmatrix} A + B^T W^{-1} B & 0 \\ 0 & W \end{bmatrix}, \tag{2.3.5}$$

and Rees *et al.* [23] have introduced augmentation block triangular preconditioners of the form

$$\mathcal{M}_k = \begin{bmatrix} A + B^T W^{-1} B & kB^T \\ 0 & W \end{bmatrix}, \tag{2.3.6}$$

where k is a scalar, $W \in \mathbb{R}^{m \times m}$ is a symmetric positive definite weight matrix. Obviously, when $k = 0$ in (2.3.6) we have $\mathcal{M}_k = \mathcal{M}_W$.

Cao [24] extended $\mathcal{M}_W$ to nonsymmetric case, i.e. for the saddle point system (2.3.1), and introduced the following augmentation block diagonal preconditioner:

$$\mathcal{D}_{Aug} = \begin{bmatrix} A + B^T W^{-1} C & 0 \\ 0 & W \end{bmatrix},$$

where $W \in \mathbb{R}^{m \times m}$ is nonsingular and such that $A + B^T W^{-1} C$ is also nonsingular.

The following theorem provides the spectrum results of the preconditioned matrix $\mathcal{D}_{Aug}^{-1}\mathcal{A}$.

Theorem 2.4 [24] *Assume that $\mathcal{A}$ is nonsingular and its (1,1) block A is singular with nullity s. Then $\lambda = 1$ is an eigenvalue of $\mathcal{D}_{Aug}^{-1}\mathcal{A}$ of geometric multiplicity n and $\lambda = -1$ is an eigenvalue of geometric multiplicity s. The remaining $m - s$ eigenvalues satisfy the relation*

$$\lambda = -\frac{\mu}{\mu + 1},$$

where μ are $m - s$ finite nonzero eigenvalues of the following generalized eigenvalue problem:

$$B^T W^{-1} C x = \mu A x.$$

Let $\{z_i\}_{i=1}^{n-m}$ be a basis of $\mathcal{N}(C)$, $\{x_i\}_{i=1}^{s}$ a basis of $\mathcal{N}(A)$, and $\{y_i\}_{i=1}^{m-s}$ a set of linearly independent vectors that complete $\mathcal{N}(C) \cup \mathcal{N}(A)$ to a basis of $\mathbb{R}^n$. Then the vectors $[z_i^T, 0^T]^T$, $i = 1, \ldots, n-m$, the vectors $[x_i^T, (W^{-1}Cx_i)^T]^T$, $i = 1, \ldots, s$, and the vectors $[y_i^T, (W^{-1}Cy_i)^T]^T$, $i = 1, \ldots, m-s$, are linearly independent eigenvectors associated with $\lambda = 1$, and the vectors $[x_i^T, -(W^{-1}Cx_i)^T]^T$, $i = 1, \ldots, s$, are linearly independent eigenvectors associated with $\lambda = -1$.

It is easy to extend $\mathcal{M}_k$ in (2.3.6) to the generalized saddle point system (2.3.1) to get

$$\mathcal{T}_{Augk+} = \begin{bmatrix} A + B^T W^{-1} C & kB^T \\ 0 & W \end{bmatrix}. \tag{2.3.7}$$

In contrast to (2.3.7), for the saddle point system (2.3.1), Cao [25] introduced augmentation block triangular preconditioners of the form

$$\mathcal{T}_{Augk-} = \begin{bmatrix} A + B^T W^{-1} C & kB^T \\ 0 & -W \end{bmatrix}, \tag{2.3.8}$$

where k is a scalar and $W \in \mathbb{R}^{m \times m}$ is nonsingular and such that $A + B^T W^{-1} C$ is invertible. We note that preconditioners in the forms (2.3.7) and (2.3.8) are different only in taking opposite sign in their (2,2) blocks. It has been shown that the preconditioners in the form (2.3.8) are more effective than those in the form (2.3.7). Especially, when $k = 2$ the augmentation block triangular preconditioner

$$\mathcal{T}_{Aug2-} = \begin{bmatrix} A + B^T W^{-1} C & 2B^T \\ 0 & -W \end{bmatrix}$$

is most effective, i.e. the preconditioned matrix $\mathcal{T}_{Aug2-}^{-1} \mathcal{A}$ has better clustering eigenvalues and makes the corresponding preconditioned Krylov subspace iterative method converges very fast.

The following theorem provides the spectrum results of the preconditioned matrix $\mathcal{T}_{Augk-}^{-1} \mathcal{A}$, $k \neq 2$.

Theorem 2.5 [25] *Assume that $\mathcal{A}$ is nonsingular and that its (1,1) block A is singular with nullity s. If $k \neq 2$, then $\lambda = 1$ is an eigenvalue of $\mathcal{T}_{Augk-}^{-1} \mathcal{A}$ of geometric multiplicity n-m, and $\lambda = \frac{k+\sqrt{k^2-4}}{2}$ and $\lambda = \frac{k-\sqrt{k^2-4}}{2}$ are two eigenvalues of geometric multiplicity s. The remaining $2(m-s)$ eigenvalues satisfy the relation*

$$\lambda = \frac{k\mu + 1 \pm \sqrt{(k\mu + 1)^2 - 4\mu(\mu + 1)}}{2(\mu + 1)},$$

where μ are the m-s finite nonzero eigenvalues of the generalized eigenvalue problem

$$B^T W^{-1} C x = \mu A x.$$

Let $\{z_i\}_{i=1}^{n-m}$ be a basis of $\mathcal{N}(C)$ and $\{x_i\}_{i=1}^{s}$ a basis of $\mathcal{N}(A)$. Then the vectors $[z_i^T, 0^T]^T$, $i = 1, \ldots, n-m$, are linearly independent eigenvectors associated with $\lambda = 1$, and the vectors $[x_i^T, \frac{-k \pm \sqrt{k^2-4}}{2}(W^{-1}Cx_i)^T]^T$, $i = 1, \ldots, s$, are linearly independent eigenvectors associated with $\lambda = \frac{k \pm \sqrt{k^2-4}}{2}$.

The following theorem providing the spectrum results of the preconditioned matrix $\mathcal{T}_{Aug2-}^{-1} \mathcal{A}$ was given.

Theorem 2.6 [25] *Assume that $\mathcal{A}$ is nonsingular and its (1,1) block A is singular with nullity s. Then $\lambda = 1$ is an eigenvalue of $\mathcal{T}_{Aug2-}^{-1} \mathcal{A}$ of multiplicity n + s associated with n linearly independent* *eigenvectors (i.e. the geometric multiplicity is n) and s linearly independent generalized eigenvectors of order 2. The remaining m-s eigenvalues satisfy the relation*

$$\lambda = \frac{\mu}{\mu + 1},$$

where μ are the m-s finite nonzero eigenvalues of the following generalized eigenvalue problem:

$$B^T W^{-1} C x = \mu A x.$$

Let $\{z_i\}_{i=1}^{n-m}$ be a basis of $\mathcal{N}(C)$, $\{x_i\}_{i=1}^{s}$ a basis of $\mathcal{N}(A)$, and $\{y_i\}_{i=1}^{m-s}$ a set of linearly independent vectors that complete $\mathcal{N}(C) \cup \mathcal{N}(A)$ to a basis of $\mathbb{R}^n$. Then the vectors $[z_i^T, 0^T]^T$, $i = 1, \ldots, n-m$, the vectors $[x_i^T, -(W^{-1} C x_i)^T]^T$, $i = 1, \ldots, s$, and the vectors $[y_i^T, -(W^{-1} C y_i)^T]^T$, $i = 1, \ldots, m-s$, are linearly independent eigenvectors associated with $\lambda = 1$.

3.3   A set of test matrices

A method has been given [24] to construct a set of generalized saddle point matrices of the form (2.3.1) such that they have high nullity. They are constructed from partitioning the matrix $\widehat{\mathcal{A}}$ in the following form:

$$\widehat{\mathcal{A}} = \begin{bmatrix} F_1 & 0 & B_u^T \\ 0 & F_2 & B_v^T \\ B_u & B_v & 0 \end{bmatrix},$$

where $\begin{bmatrix} F_1 & 0 \\ 0 & F_2 \end{bmatrix} \equiv A$ is positive real, i.e. $A + A^T$ is symmetric positive definite. The matrix $\widehat{\mathcal{A}}$ arises from the discretization by the "maker-and-cell" (MAC) finite difference scheme [15] of a "leaky" two-dimensional lid-driven cavity problem in a square domain $(0 \le x \le 1 : 0 \le y \le 1)$. Then the matrices $[B_u, B_v]$ and $[B_u, B_v]^T$ are replaced by a random matrix $\widehat{C}$ with the same sparsity as $[B_u, B_v]$ and a random matrix $\widehat{B}^T$ with the same sparsity as $[B_u, B_v]^T$, respectively. Furthermore, $\widehat{C}(1 : m, 1 : m)$ and $\widehat{B}(1 : m, 1 : m)$ are respectively replaced by $C1 = \widehat{C}(1 : m, 1 : m) - \frac{3}{2} I_m$ and $B_1 = \widehat{B}(1 : m, 1 : m) - \frac{3}{2} I_m$, such that C_1 and B_1 are nonsingular. Denote $C_2 = \widehat{C}(1 : m, m+1 : n)$ and $B_2 = \widehat{B}(1 : m, m+1 : n)$, then we have $C = [C_1, C_2]$ and $B = [B_1, B_2]$ with $B_1, C_1 \in \mathbb{R}^{m \times m}$ and $B_2, C_2 \in \mathbb{R}^{m \times (n-m)}$. Obviously, the resulting generalized saddle point matrix

$$\mathcal{A} = \begin{bmatrix} A & B^T \\ C & 0 \end{bmatrix} \tag{2.3.9}$$

satisfies

$$rank(C) = rank(B) = m.$$

From the matrix $\mathcal{A}$ in (2.3.9) we construct the following four generalized saddle point matrices:

$$\mathcal{A}_i = \begin{bmatrix} A_i & B^T \\ C & 0 \end{bmatrix}, \quad i = 1, \ldots, 4,$$

where A_i is constructed from A by making its first $i \times \frac{m}{4}$ rows and columns with zero entries. Note that A_i is semi-positive real and its nullity is $i \times \frac{m}{4}$.

Acknowledgments

This work is supported by NSFC Project 10871051.

Bibliography

[1] Benzi M, Golub GH, Liesen J. Numerical solution of saddle point problems. Acta Numerica 2005; 14: 1-137.

[2] Murphy MF, Golub GH, Wathen AJ. A note on preconditioning for indefinite linear systems. SIAM J Sci Comput 2000; 21(6): 1969-72.

[3] Ipsen ICF. A note on preconditioning nonsymmetric matrices. SIAM J Sci Comput 2001; 23(3): 1050-1.

[4] de Sturler E, Liesen J. Block diagonal and constraint preconditioners for nonsymmetric indefinite linear systems. Part I: Theory. SIAM J Sci Comput 2005; 26(5): 1598-619.

[5] Stewart GW, Sun JG. Matrix perturbation theory. Boston: Academic Press 1990.

[6] Cao ZH. A note on block diagonal and constraint preconditioners for non-symmetric indefinite linear systems. Inter J Computer Math 2006; 83(4): 383-95.

[7] Siefert C, de Sturler E. Preconditioners for generalized saddle-point problems. SIAM J Numer Anal 2006; 44(3): 1275-96.

[8] Cao ZH. Block triangular Schur complement preconditioners for saddle point problems and application to the Oseen equations. Applied Numer Math 2010; 60(3): 193-207.

[9] Keller C, Gould NIM, Wathen AJ. Constraint preconditioning for indefinite linear systems. SIAM J Matrix Anal Appl 2000; 21(4): 1300-17.

[10] Cao ZH. A note on constraint preconditioning for nonsymmetric indefinite matrices. SIAM J Matrix Anal Appl 2002; 24(1): 121-5.

[11] Dollar HS. Constraint-style preconditioners for regularized saddle point problems. SIAM J Matrix Anal Appl 2007; 29(2): 672-84.

[12] Cao ZH. Constraint Schur complement preconditioners for nonsymmetric saddle point problems. Applied Numer Math 2009; 59(1): 151-69.

[13] Cao ZH. A note on spectrum distribution of constraint preconditioned generalized saddle point matrices. Numer Linear Algebra Appl 2009; 16(6): 503-16.

[14] Saad Y. Iterative methods for sparse linear systems. Boston: PWS Publishing Company 1996.

[15] Elman HC. Preconditioning for the steady-state Navier-Stokes equations with low viscosity. SIAM J Sci Comput 1999; 20(4): 1299-316.

[16] Elman HC, Silvester DJ, Wathen AJ. Finite elements and fast iterative solvers. New York: Oxford University Press 2005.

[17] Elman HC, Howle VE, Shadid J et al. Block preconditioners based on approximate commutators. SIAM J Sci Comput 2006: 27(5): 1651-68.

[18] Silvester DJ, Kechkar N. Stabilized bilinear-constant velocity-pressure finite elements for the conjugate gradient solution of the Stokes problem. Comput Methods Appl Mech Engrg 1990; 79(1): 71-86.

[19] Silvester DJ, Wathen AJ. Fast iterative solution of stabilized Stokes systems II: Using general block preconditioners. SIAM J Numer Anal 1994; 31(5): 1352-67.

[20] Golub GH, Greif C, Varah JM. An algebraic analysis of a block diagonal preconditioners for saddle point systems. SIAM J Matrix Anal Appl 2006; 27(3): 779-92.

[21] Cao ZH. A note on spectrum analysis of augmentation block triangular Schur complement preconditioners. 2009; unpublished manuscript.

[22] Greif C, Schötzau D. Preconditioners for saddle point linear systems with highly singular (1,1) blocks. Electronic Transactions on Numerical Analysis 2006; 22: 114-21.

[23] Rees T, Greif C. A preconditioner for linear systems arising from interior optimization methods. SIAM J Sci Comput 2007; 29(5): 1992-2007.

[24] Cao ZH. Augmentation block preconditioners for saddle-point matrices with singular (1,1) blocks. Numer Linear Algebra Appl 2008; 15(6): 515-33.

[25] Cao ZH. A note on preconditioning for generalized saddle point matrices with singular (1,1) blocks. 2009; unpublished manuscript.

Chapter 3

Block Preconditioners for Saddle Point Problems Resulting from Discretizations of Partial Differential Equations

Piotr Krzyżanowski[1]

Abstract: We discuss a family of block preconditioners for iterative solution of symmetric saddle point type problems arising from PDE discretizations. The building blocks consist of preconditioners for smaller sized, symmetric positive definite operators, which induce a norm in which the whole system is continuous and stable uniformly with respect to the mesh size h. We provide eigenvalue estimates and derive conditions under which the conjugate residual method using block preconditioners has convergence rate bounded independently of h.

Keywords: *saddle point systems, partial differential equations, discretizations, stability, iterative solution, conjugate residual method, block preconditioners, robust*

1　Introduction

In this chapter we consider a symmetric system of linear equations with block structure,

$$\mathcal{M} \begin{pmatrix} u \\ p \end{pmatrix} \equiv \begin{pmatrix} A & B^T \\ B & -C \end{pmatrix} \begin{pmatrix} u \\ p \end{pmatrix} = \begin{pmatrix} F \\ G \end{pmatrix}. \tag{1.1}$$

Many such systems arise from the discretization of (system of) partial differential equations. For example, Stokes equations discretized with finite elements [1], or a mixed finite element method for second order elliptic PDEs [2] lead to a positive definite matrix A and $C = 0$, so that (1.1) has a genuine saddle point structure. Certain other PDE problems – such as the time harmonic Maxwell equations [3] – may result in indefinite A, or in semidefinite A with large kernel, which gives (1.1) the structure of a so called generalized saddle point problem. Linear elasticity equations of nearly incompressible materials discretized with mixed finite elements result in both A and C positive definite [4, 5], having

---

[1]Institute of Applied Mathematics, University of Warsaw, Poland;　e-mail: piotr.krzyzanowski@mimuw.edu.pl

Owe Axelsson and János Karátson (Eds)

thus a nature of a penalized saddle point problem. All systems mentioned above have a common feature that (1.1) is indefinite. Let us remark that e.g. substructuring methods for elliptic PDEs [6] also lead to a block system (1.1), but then with $\mathcal{M}$ positive definite.

The specific structure of (1.1) makes it possible to design efficient solution methods which intensively exploit the properties of the system, see the recent survey [7] on the state-of-the-art in this field. Systems derived from the discretization of PDEs are usually very large and sparse, and typically are solved by some iterative method. Unfortunately, these systems are ill-conditioned with respect to the mesh size h, so preconditioning is necessary in order to keep the number of iterations within a reasonable limit. Applying a left preconditioner $\mathcal{P}$, one then solves a problem with a preconditioned matrix $\mathcal{P}^{-1}\mathcal{M}$.

Assume first that A is nonsingular. Then the following dual decomposition of the block system matrix [8] is valid:

$$\begin{pmatrix} A & E^T \\ F & -C \end{pmatrix} = \begin{pmatrix} I & \\ FA^{-1} & I \end{pmatrix} \begin{pmatrix} A & \\ & S_A \end{pmatrix} \begin{pmatrix} I & A^{-1}E^T \\ & I \end{pmatrix}. \tag{1.2}$$

Here $S_A = -C - FA^{-1}E^T$ is the Schur complement with respect to A. Analogously, when C is nonsingular, one gets another, primal decomposition:

$$\begin{pmatrix} A & E^T \\ F & -C \end{pmatrix} = \begin{pmatrix} I & -E^T C^{-1} \\ & I \end{pmatrix} \begin{pmatrix} S_C & \\ & -C \end{pmatrix} \begin{pmatrix} I & \\ -C^{-1}F & I \end{pmatrix}, \tag{1.3}$$

with the Schur complement with respect to the $(2,2)$ block of $\mathcal{M}$ being $S_C = A + E^T C^{-1} F$. The aim of this chapter is to discuss properties of block preconditioners inspired by the above decompositions. We shall consider preconditioners of the form

$$\mathcal{P}_{\mathrm{d}} = \begin{pmatrix} I & \\ cBA_0^{-1} & I \end{pmatrix} \begin{pmatrix} A_0 & \\ & S_0 \end{pmatrix} \begin{pmatrix} I & dA_0^{-1}B^T \\ & I \end{pmatrix} \tag{1.4}$$

or

$$\mathcal{P}_{\mathrm{p}} = \begin{pmatrix} I & dB^T S_0^{-1} \\ & I \end{pmatrix} \begin{pmatrix} A_0 & \\ & S_0 \end{pmatrix} \begin{pmatrix} I & \\ cS_0^{-1}B & I \end{pmatrix}, \tag{1.5}$$

where A_0 and S_0 are symmetric, positive (or negative) definite matrices and c, d are prescribed real numbers. By analogy to (1.2) and (1.3), and in accordance with [9], we will refer to $\mathcal{P}_{\mathrm{d}}$ as the family of dual block preconditioners and to $\mathcal{P}_{\mathrm{p}}$ as the family of primal block preconditioners.

Many popular block preconditioners can be formed by choosing appropriate values of c and d in formulas above. For example, a block diagonal preconditioner [5, 10–14] corresponds to $c = d = 0$ above. Block triangular preconditioners [15–17] and the Bramble–Pasciak preconditioner [18] as well, are obtained with either c or d equal to zero. The choice $c = d = 1$ in (1.4) produces the symmetric indefinite preconditioner [8, 19–21], while the same choice in (1.5) leads to the primal based penalty preconditioner [9, 22]. Let us note in passing that symmetric indefinite preconditioners have been used for a long time in substructuring methods for second order elliptic equations while dealing with inexact subdomain solvers [6].

Block preconditioned Uzawa-type methods have also attracted research interest [23, 24]. An augmented Lagrangian technique, which replaces the original block A with $A + B^T W B$ for some carefully chosen symmetric positive definite matrix W, has been analyzed in [25] and since then has become popular, see e.g. [26–28].

Certain Krylov space methods which rely on the symmetry of the problem (e.g. the conjugate gradient method or the conjugate residuals method [29]) need a custom inner product matrix $\mathcal{H}$ in order to symmetrize the preconditioned matrix. Therefore, in many works on this subject, such proper matrices are provided and properties of the resulting systems are analyzed [5, 15, 18, 20, 21, 30]. Stoll *et al.* [31, 32] extended the approach of Bramble and Pasciak and made it possible to analyze a broader range of preconditioners. This idea has then been further developed for dual block preconditioners in [33] and the present chapter summarizes and extends these results.

The above list of literature related to block preconditioners and their properties is by no means exhaustive, see e.g. the survey [7] for more references.

Let us stress that when (1.1) arises from a finite element discretization of PDEs, there is a possibility to use other than block preconditioning approaches. The inner nature of the underlying PDE problem may make it possible to use *optimal* multigrid [34, 35] solvers or to construct specific and *extremely* efficient preconditioners based on domain decomposition [6, 36] techniques. For example, for an iterative substructuring method for Stokes equations discretized with certain finite elements, a number of highly scalable parallel preconditioners based on domain decomposition: balancing Neumann–Neumann [37], BDDC [38] and FETI-DP [39] has recently been developed and proved (almost) optimal with respect to the mesh size h. Widlund and coworkers have also obtained similar results for mixed approximations of linear elasticity. Similarly, for the multigrid, there are several optimal results for specific PDEs and discretizations which lead to saddle point type problems, e.g. [40–43] to mention just a few, but still many other PDEs or their discretizations are not covered by existing theory.

On the other hand, one has to remember that for many types of discretizations and problems, specialized methods based on the direct construction of a multigrid or domain decomposition preconditioner – although usually outperforming block preconditioners [44] – may take a considerable effort to develop, implement and analyse. This stands in contrast to the simpler case of elliptic second order equations, where both the theory and practice of fast solution or preconditioning methods is very mature. Since the block preconditioning approach as discussed in this chapter turns out to be based on preconditioners for symmetric positive definite matrices, this property makes it a viable and robust alternative to custom methods, as in this case one can efficiently utilize existing theory and software to solve more complex problems.

Some approaches to block diagonal preconditioning assume exact solution (or almost exact, using for example a multigrid solver) of e.g. one of the diagonal blocks of (1.1) or its Schur complement, see [45] for examples from the computational fluid dynamics. While our approach does not exclude such a treatment, it also allows the use of other, spectrally equivalent preconditioning blocks, providing more flexibility in the choice of A_0 and S_0. A general construction methodology, which leads to good candidates for A_0 and S_0, is discussed in Chapter 4 of this volume.

The rest of the chapter is organized as follows. First, in Section 2, we introduce basic requirements necessary to apply either primal or dual block preconditioners in the Preconditioned Conjugate Residuals (PCR) method [29] and then we discuss some implementation issues. In Section 3, we present an algebraic spectral analysis of dual and primal block preconditioners, which can be used also outside the PDE context. Next, in Section 4 we provide some more detailed conditions for certain preconditioners encountered in the literature, while in Section 5 we apply the results in a variational framework well suited to PDE applications, showing optimality conditions for preconditioners under consideration in three settings: generalized, penalized or stabilized saddle point problems.

2   Block Preconditioners and the PCR Method

Throughout the chapter we shall assume that

- $|c|, |d| \leq 1$,

- $\mathcal{M}$, A_0 and S_0 are real symmetric and nonsingular,

- there exist $\varepsilon_1, \varepsilon_2 \in \{-1, +1\}$ such that $\tilde{A}_0 = \varepsilon_1 A_0$ and $\tilde{S}_0 = \varepsilon_2 S_0$ are positive definite,

- the inverses of A_0 and S_0 are *easy to apply* to a vector.

The third assumption simply means that A_0 or S_0 are either positive or negative definite. We will need $\tilde{A}_0$ and $\tilde{S}_0$ in Section 3 to define an auxiliary inner product matrix.

Let us note some relationship between $\mathcal{P}_p$ and $\mathcal{P}_d$. First, the two preconditioners are identical when $cd = 0$, leading to block diagonal or block triangular matrices. Next, $\mathcal{P}_p$ is essentially $\mathcal{P}_d$ with the unknowns and the equations swapped. However, because the properties of the diagonal blocks in (1.1) are in general different, $\mathcal{P}_p$ and $\mathcal{P}_d$ present two real alternatives to preconditioning (1.1). Direct computation shows that

$$\mathcal{P}_d^{-1} = \underbrace{\begin{pmatrix} I & -dA_0^{-1}B^T \\ & I \end{pmatrix}}_{\mathcal{U}_d} \begin{pmatrix} A_0^{-1} & \\ & S_0^{-1} \end{pmatrix} \underbrace{\begin{pmatrix} I & \\ -cBA_0^{-1} & I \end{pmatrix}}_{\mathcal{L}_d} \tag{2.1}$$

and analogously

$$\mathcal{P}_p^{-1} = \underbrace{\begin{pmatrix} I & \\ -cS_0^{-1}B & I \end{pmatrix}}_{\mathcal{L}_p} \begin{pmatrix} A_0^{-1} & \\ & S_0^{-1} \end{pmatrix} \underbrace{\begin{pmatrix} I & -dB^T S_0^{-1} \\ & I \end{pmatrix}}_{\mathcal{U}_p}. \tag{2.2}$$

Therefore, solving a system with $\mathcal{P}_d$ requires one solve with S_0 and at most two solves with A_0, cf. Algorithm 2, while applying $\mathcal{P}_p$ to a vector takes one solve with A_0 and at most two solves with S_0 (Algorithm 3). In consequence, the costs of applying $\mathcal{P}_d^{-1}$ and $\mathcal{P}_p^{-1}$ to a vector may differ when $cd \neq 0$; when $cd = 0$, both types of preconditioners require only one solve with A_0 and one with S_0.

The conjugate residual method requires a matrix which is symmetric with respect to a certain inner product defined by a symmetric positive definite matrix $\mathcal{H}$. It is an iterative method, which minimizes certain error norm over a Krylov space. It can be implemented using a generic algorithm, Odir [29], which is based on a three term recursion. The following proposition and its corollary, whose proofs are straightforward, provide good choices for both types of preconditioners, cf. [9, 20, 33].

Proposition 3.1. *Let* $\delta \in \{-1, +1\}$.

  (i) *If the block diagonal matrix* $\mathcal{H}_d$ *is of the form*

$$\mathcal{H}_d = \delta \begin{pmatrix} A_0 - cA & \\ & S_0 + cdBA_0^{-1}B^T + dC \end{pmatrix}, \tag{2.3}$$

    then $\mathcal{H}_d \mathcal{P}_d^{-1} \mathcal{M}$ *is symmetric.*

  (ii) *If the block diagonal matrix* $\mathcal{H}_p$ *is of the form*

$$\mathcal{H}_p = \delta \begin{pmatrix} A_0 + cdB^T S_0^{-1}B - cA & \\ & S_0 + dC \end{pmatrix}, \tag{2.4}$$

    then $\mathcal{H}_p \mathcal{P}_p^{-1} \mathcal{M}$ *is symmetric.*

Corollary 3.1. *Let $\mathcal{H}_d$ and $\mathcal{H}_p$ be defined as in Proposition 3.1.*

(i) Let

$$Q_d = \mathcal{M} \mathcal{P}_d^{-T} \mathcal{H}_d \mathcal{P}_d^{-1} \mathcal{M}. \tag{2.5}$$

If $\mathcal{H}_d$ is positive definite, then systems (1.1) and

$$\mathcal{H}_d \mathcal{P}_d^{-1} \mathcal{M} \begin{pmatrix} u \\ p \end{pmatrix} = \mathcal{H}_d \mathcal{P}_d^{-1} \begin{pmatrix} f \\ g \end{pmatrix}$$

are equivalent, Q_d is symmetric positive definite and

$$\mathrm{Odir}(Q_d, \mathcal{H}_d^{-1}, \mathcal{H}_d \mathcal{P}_d^{-1} \mathcal{M}),$$

cf. [29], defines a PCR method which can be used to solve (1.1).

(ii) Let

$$Q_p = \mathcal{M} \mathcal{P}_p^{-T} \mathcal{H}_p \mathcal{P}_p^{-1} \mathcal{M}. \tag{2.6}$$

If $\mathcal{H}_p$ is positive definite, then systems (1.1) and

$$\mathcal{H}_p \mathcal{P}_p^{-1} \mathcal{M} \begin{pmatrix} u \\ p \end{pmatrix} = \mathcal{H}_p \mathcal{P}_p^{-1} \begin{pmatrix} f \\ g \end{pmatrix}$$

are equivalent, Q_p is symmetric positive definite and

$$\mathrm{Odir}(Q_p, \mathcal{H}_p^{-1}, \mathcal{H}_p \mathcal{P}_p^{-1} \mathcal{M})$$

defines a PCR method which can be used to solve (1.1).

If $\mathcal{H} \mathcal{P}^{-1} \mathcal{M}$ (where the pair $\mathcal{H}, \mathcal{P}$ stands for $\mathcal{H}_d, \mathcal{P}_d$ in the dual case and for $\mathcal{H}_p, \mathcal{P}_p$ in the primal case) is symmetric and positive definite, one can switch to a Conjugate Gradient method. Many works indeed aim towards a construction of such $\mathcal{P}$ and $\mathcal{H}$ so that $\mathcal{H} \mathcal{P}^{-1} \mathcal{M}$ is positive definite, [9, 18, 20, 21]. Such an approach, however, requires some careful scaling of the preconditioning blocks A_0 or S_0.

It is well-known [29, 46] that using a few additional vectors one can implement the algorithm $\mathrm{Odir}(Q, \mathcal{H}^{-1}, \mathcal{H} \mathcal{P}^{-1} \mathcal{M})$ in such a way that only one solve with the preconditioner $\mathcal{P}$ is required per iteration, see Algorithm 1. Moreover, the two required matrix-vector multiplications: $\mathcal{P}^{-1}(\mathcal{M} z_i)$ and $\mathcal{H} \mathcal{P}^{-1}(\mathcal{M} z_i)$, can be achieved with only one solve with $\mathcal{P}$ and several multiplications with A, B, C as in Algorithm 2 or Algorithm 3 (notice again that the primal preconditioner is essentially the dual one with the equations and the unknowns swapped). Direct reference to A_0 and S_0 can thus be completely avoided, a desirable feature especially when A_0 and S_0 are only available through their inverses.

3 Convergence Analysis

Here we extend the results of [33] and provide eigenvalue estimates in terms of certain "natural" properties of the system $\mathcal{M}$ and the preconditioning blocks A_0, S_0. These estimates, valid for both dual and primal block preconditioners, are then used to provide a convergence rate bound for the PCR method. Later, in Section 5 we show how the algebraic results of the present section can be applied in the context of PDE discretizations.

Algorithm 1: PCR algorithm to solve $\mathcal{M}x = y$ with preconditioner $\mathcal{P}$ and inner product $\mathcal{H}$, see [29].

$p_{-1} = z_{-1} = q_{-1} = 0;\, m_{-1} = 1;$
$r_0 = \mathcal{P}^{-1}(y - \mathcal{M}x_0);$
$p_0 = r_0;\, z_0 = \mathcal{P}^{-1}\mathcal{M}p_0;\, q_0 = \mathcal{H}\mathcal{P}^{-1}\mathcal{M}p_0;$
$i = 0;$
while *not converged* **do**

$\quad m_i = q_i^T z_i;$

$\quad \alpha_i = \dfrac{q_i^T r_i}{m_i};$

$\quad x_{i+1} = x_i + \alpha_i p_i;$

$\quad r_{i+1} = r_i - \alpha_i z_i;$

$\quad$ **begin** /* see Algorithms 2 or 3 for details */

$\quad\quad s_{i+1} = \mathcal{P}^{-1}\mathcal{M}z_i;$

$\quad\quad w_{i+1} = \mathcal{H}\mathcal{P}^{-1}\mathcal{M}z_i;$

$\quad$ **end**

$\quad \gamma_i = \dfrac{z_i^T w_{i+1}}{m_i};$

$\quad \sigma_i = \dfrac{z_{i-1}^T w_{i+1}}{m_{i-1}};$

$\quad p_{i+1} = z_i - \gamma_i p_i - \sigma_i p_{i-1};$

$\quad z_{i+1} = s_{i+1} - \gamma_i z_i - \sigma_i z_{i-1};$

$\quad q_{i+1} = w_{i+1} - \gamma_i q_i - \sigma_i q_{i-1};$

$\quad i = i + 1;$

end

Algorithm 2: Calculation of $\mathcal{P}_{\mathrm{d}}^{-1}\mathcal{M}z$ and $\mathcal{H}_{\mathrm{d}}\mathcal{P}_{\mathrm{d}}^{-1}\mathcal{M}z$.

$\begin{pmatrix} u \\ p \end{pmatrix} = \mathcal{M}z;$

begin /* First compute $\begin{pmatrix} x \\ y \end{pmatrix} = \mathcal{P}_{\mathrm{d}}^{-1}\begin{pmatrix} u \\ p \end{pmatrix} = \mathcal{P}_{\mathrm{d}}^{-1}\mathcal{M}z.$ */

$\quad w = A_0^{-1}u;$

$\quad r = p - cBw;$

$\quad y = S_0^{-1}r;$

$\quad q = dB^T y;$

$\quad v = A_0^{-1}q;$

$\quad x = w - v;$

end

begin /* Next compute $\begin{pmatrix} s \\ t \end{pmatrix} = \mathcal{H}_{\mathrm{d}}\mathcal{P}_{\mathrm{d}}^{-1}\begin{pmatrix} u \\ p \end{pmatrix} = \mathcal{H}_{\mathrm{d}}\mathcal{P}_{\mathrm{d}}^{-1}\mathcal{M}z$, reusing previously computed vectors. */

$\quad s = \delta(u - q - cAx);$

$\quad t = \delta(r + dCy + cBv);$

end

Algorithm 3: Calculation of $\mathcal{P}_p^{-1}\mathcal{M}z$ and $\mathcal{H}_p\mathcal{P}_p^{-1}\mathcal{M}z$.

$\begin{pmatrix} u \\ p \end{pmatrix} = \mathcal{M}z$;

begin /* First compute $\begin{pmatrix} x \\ y \end{pmatrix} = \mathcal{P}_p^{-1}\begin{pmatrix} u \\ p \end{pmatrix} = \mathcal{P}_p^{-1}\mathcal{M}z$. */

$\quad w = S_0^{-1}p$;
$\quad r = u - dB^T w$;
$\quad x = A_0^{-1}r$;
$\quad q = cBx$;
$\quad v = S_0^{-1}q$;
$\quad y = w - v$;

end

begin /* Next compute $\begin{pmatrix} s \\ t \end{pmatrix} = \mathcal{H}_p\mathcal{P}_p^{-1}\begin{pmatrix} u \\ p \end{pmatrix} = \mathcal{H}_p\mathcal{P}_p^{-1}\mathcal{M}z$, reusing previously computed vectors. */

$\quad s = \delta(r + dB^T v - cAx)$;
$\quad t = \delta(p - cBx + dCy)$;

end

As we are going to analyze both primal and dual versions of the preconditioners, we will, where possible, use the "generic" triple Q, $\mathcal{H}$, $\mathcal{P}$ to denote either a triple of Q_d, $\mathcal{H}_d$, $\mathcal{P}_d$ in the case of dual block preconditioners family or Q_p, $\mathcal{H}_p$, $\mathcal{P}_p$ for primal block preconditioners.

In what follows, we shall refer to several vector norms in R^d. For a given vector $x \in R^d$, we denote by $\|x\| = (x^T x)^{1/2}$ its usual Euclidean norm, while for a positive definite matrix X we shall denote the energy norm induced by X by $\|x\|_X = (x^T X x)^{1/2}$. We shall use the same symbols while referring to matrix norms induced by the above vector norms.

3.1   Basic assumptions

We assume that A is an $n \times n$, while C is an $m \times m$ matrix. Inspired by the block nature of the problem, which imposes a decomposition of the unknowns into two parts $(u, p) \in R^n \times R^m$, let us define a block diagonal, symmetric, positive definite matrix

$$\mathcal{J} = \begin{pmatrix} \tilde{A}_0 & \\ & \tilde{S}_0 \end{pmatrix},$$

which gives rise to a norm

$$\left\| \begin{pmatrix} u \\ p \end{pmatrix} \right\|_{\mathcal{J}}^2 = \|u\|_{\tilde{A}_0}^2 + \|p\|_{\tilde{S}_0}^2,$$

and assume there exist positive constants m_0 and m_1 such that

$$m_0\|x\|_{\mathcal{J}} \leq \|\mathcal{M}x\|_{\mathcal{J}^{-1}} \leq m_1\|x\|_{\mathcal{J}} \qquad \forall x \in R^n \times R^m. \tag{3.1}$$

At the same time we suppose that there exists a constant $b_0 > 0$ such that for any $u \in R^n$ and $p \in R^m$,

$$|p^T B u| \leq b_0\|u\|_{\tilde{A}_0}\|p\|_{\tilde{S}_0}. \tag{3.2}$$

Finally, we assume that for some $\delta \in \{-1, +1\}$, $\mathcal{H}$ defined in Proposition 3.1 (equal to either $\mathcal{H}_d$ or $\mathcal{H}_p$) is positive definite and is spectrally equivalent to $\mathcal{J}$ with positive constants h_0 and h_1:

$$h_0 \|x\|_{\mathcal{H}} \leq \|x\|_{\mathcal{J}} \leq h_1 \|x\|_{\mathcal{H}}, \qquad \forall x \in R^n \times R^m. \tag{3.3}$$

Note that condition (3.3) leads to the assumption in Corollary 3.1.

The following proposition also shows that (3.1) and (3.3) are not quite independent: the lower bound in (3.3) can actually be derived from the upper bound in (3.1).

Proposition 3.2. *If (3.1) holds, then there exists $h_0 > 0$, dependent only on m_1, such that*

$$h_0 \|x\|_{\mathcal{H}} \leq \|x\|_{\mathcal{J}} \qquad \forall x \in R^n \times R^m.$$

Proof. Indeed, it follows from (3.1) that for any $x = (u, p)$

$$\|Au + B^T p\|_{\tilde{A}_0^{-1}}^2 + \|Bu - Cp\|_{\tilde{S}_0^{-1}}^2 \leq m_1^2 (\|u\|_{\tilde{A}_0}^2 + \|p\|_{\tilde{S}_0}^2),$$

so in particular, taking x equal to either $(u, 0)$ or $(0, p)$, we get

$$\begin{aligned} \|B^T p\|_{\tilde{A}_0^{-1}} &\leq m_1 \|p\|_{\tilde{S}_0}, & \|Bu\|_{\tilde{S}_0^{-1}} &\leq m_1 \|u\|_{\tilde{A}_0}, \\ \|Au\|_{\tilde{A}_0^{-1}} &\leq m_1 \|u\|_{\tilde{A}_0}, & \|Cp\|_{\tilde{S}_0^{-1}} &\leq m_1 \|u\|_{\tilde{S}_0}. \end{aligned} \tag{3.4}$$

Because for any matrices X and $\tilde{X}_0$, with $\tilde{X}_0$ symmetric positive definite, we have

$$\sup_{x \neq 0} \frac{x^T X x}{\|x\|_{\tilde{X}_0}^2} \leq \sup_{x \neq 0} \frac{\|\tilde{X}_0^{-1/2} X x\| \|\tilde{X}_0^{1/2} x\|}{\|\tilde{X}_0^{1/2} x\|^2} = \sup_{x \neq 0} \frac{\|X x\|_{\tilde{X}_0^{-1}}}{\|x\|_{\tilde{X}_0}},$$

the last two inequalities in (3.4) yield additionally

$$u^T A u \leq m_1 \|u\|_{\tilde{A}_0}^2 \qquad \text{and} \qquad p^T C p \leq m_1 \|p\|_{\tilde{S}_0}^2 \tag{3.5}$$

for all $u \in R^n$ and $p \in R^m$.

Now, since for any $\mathcal{H}$ as in Proposition 3.1 and $x = (u, p) \in R^n \times R^m$ it holds

$$\|x\|_{\mathcal{H}}^2 = x^T \mathcal{H} x \leq \|u\|_{\tilde{A}_0}^2 + \|Bu\|_{\tilde{S}_0^{-1}}^2 + |u^T A u| + \|B^T p\|_{\tilde{A}_0^{-1}}^2 + \|p\|_{\tilde{S}_0}^2 + |p^T C p|,$$

we conclude from (3.4) and (3.5) that

$$\|x\|_{\mathcal{H}}^2 \leq 2(m_1^2 + m_1 + 1) \|x\|_{\mathcal{J}}^2 \qquad \forall x \in R^n \times R^m.$$

Taking into account the specific form of $\mathcal{H}_d$ or $\mathcal{H}_p$ we can see that the factor 2 can actually be removed from the above bound. ∎

3.2 Eigenvalue estimates

Lemma 3.1. *Under assumptions of Section 3.1, the following estimates hold:*

$$\|B^T S_0^{-1} p\|_{\tilde{A}_0^{-1}} \leq b_0 \|p\|_{\tilde{S}_0^{-1}} \qquad \forall p \in R^m \tag{3.6}$$

and

$$\|Bu\|_{\tilde{S}_0^{-1}} \leq b_0 \|u\|_{\tilde{A}_0} \qquad \forall u \in R^n. \tag{3.7}$$

Proof. Since

$$\|B^T S_0^{-1} p\|_{\tilde{A}_0^{-1}} = \|B^T \tilde{S}_0^{-1} p\|_{\tilde{A}_0^{-1}} = \|\tilde{A}_0^{-1/2} B^T \tilde{S}_0^{-1/2} \tilde{S}_0^{-1/2} p\| \le \|\tilde{A}_0^{-1/2} B^T \tilde{S}_0^{-1/2}\| \|p\|_{\tilde{S}_0^{-1}}$$

and similarly

$$\|Bu\|_{\tilde{S}_0^{-1}} = \|\tilde{S}_0^{-1/2} B \tilde{A}_0^{-1/2} \tilde{A}_0^{1/2} u\| \le \|\tilde{S}_0^{-1/2} B \tilde{A}_0^{-1/2}\| \|u\|_{\tilde{A}_0},$$

it only remains to estimate $\|\tilde{S}_0^{-1/2} B \tilde{A}_0^{-1/2}\|$. From assumption (3.2) it follows that for any $q \in R^m$ and $v \in R^n$,

$$|q^T \tilde{S}_0^{-1/2} B \tilde{A}_0^{-1/2} v| \le b_0 \|\tilde{A}_0^{-1/2} v\|_{\tilde{A}_0} \|\tilde{S}_0^{-1/2} q\|_{\tilde{S}_0} = b_0 \|v\| \|q\|$$

so that $\|\tilde{S}_0^{-1/2} B \tilde{A}_0^{-1/2}\| \le b_0$, which ends the proof. ∎

Theorem 3.1. *Suppose the assumptions of Section 3.1 are fulfilled. If λ is an eigenvalue of $\mathcal{P}_d^{-1} \mathcal{M}$ or of $\mathcal{P}_p^{-1} \mathcal{M}$, then it is real and satisfies*

$$\frac{m_0}{2(1+b_0^2)} \le |\lambda| \le 2m_1(1+b_0^2).$$

We will conduct the proof only for the primal block preconditioner; the proof in the case of dual block preconditioner is completely analogous. Let us start with two lemmas. Recall from (2.2) that $\mathcal{P}_p^{-1} = \mathcal{L}_p \mathcal{J}^{-1} I \mathcal{U}_p$, where

$$I = \begin{pmatrix} \varepsilon_1 I & \\ & \varepsilon_2 I \end{pmatrix}.$$

Lemma 3.2. *Under assumptions of Section 3.1,*

$$\frac{1}{2(1+b_0^2)} x^T \mathcal{J} x \le x^T \mathcal{L}_p{}^T \mathcal{J} \mathcal{L}_p x \le 2(1+b_0^2) x^T \mathcal{J} x$$

for all $x \in R^n \times R^m$.

Proof. Setting $x = \begin{pmatrix} u \\ p \end{pmatrix}$, where $u \in R^n$, $p \in R^m$, we have

$$x^T \mathcal{L}_p{}^T \mathcal{J} \mathcal{L}_p x = \|u\|_{\tilde{A}_0}^2 + \|p - c S_0^{-1} Bu\|_{\tilde{S}_0}^2.$$

Estimating from above we obtain, by means of Lemma 3.1,

$$\begin{aligned}
x^T \mathcal{L}_p{}^T \mathcal{J} \mathcal{L}_p x &\le \|u\|_{\tilde{A}_0}^2 + 2\|p\|_{\tilde{S}_0}^2 + 2\|S_0^{-1} Bu\|_{\tilde{S}_0}^2 \\
&= \|u\|_{\tilde{A}_0}^2 + 2\|p\|_{\tilde{S}_0}^2 + 2\|Bu\|_{\tilde{S}_0^{-1}}^2 \\
&\le 2(1+b_0^2)(\|u\|_{\tilde{A}_0}^2 + \|p\|_{\tilde{S}_0}^2).
\end{aligned}$$

On the other hand, employing Lemma 3.1 again,

$$\|p\|_{\tilde{S}_0}^2 = \|p - c S_0^{-1} Bu + c S_0^{-1} Bu\|_{\tilde{S}_0}^2 \le 2\|p - c S_0^{-1} Bu\|_{\tilde{S}_0}^2 + 2b_0\|u\|_{\tilde{A}_0}^2,$$

so we conclude that also

$$\|u\|_{\tilde{A}_0}^2 + \|p\|_{\tilde{S}_0}^2 \le 2(1+b_0^2)(\|u\|_{\tilde{A}_0}^2 + \|p - c S_0^{-1} Bu\|_{\tilde{S}_0}^2) = 2(1+b_0^2) x^T \mathcal{L}_p{}^T \mathcal{J} \mathcal{L}_p x. \quad ∎$$

Lemma 3.3. *Under assumptions of Section 3.1,*

$$\frac{1}{2(1+b_0^2)}x^T \mathcal{J}^{-1}x \le x^T \mathcal{U}_p^T I^T \mathcal{J}^{-1} I \mathcal{U}_p x \le 2(1+b_0^2)x^T \mathcal{J}^{-1}x$$

for all $x \in R^n \times R^m$.

Proof. Let us begin by noticing that for any y, $y^T I^T \mathcal{J}^{-1} I y = y^T \mathcal{J}^{-1} y$. Therefore for $x = \begin{pmatrix} u \\ p \end{pmatrix}$ it holds

$$x^T \mathcal{U}_p^T I^T \mathcal{J}^{-1} I \mathcal{U}_p x = x^T \mathcal{U}_p^T \mathcal{J}^{-1} \mathcal{U}_p x = \|u - dB^T S_0^{-1} p\|_{\tilde{A}_0^{-1}}^2 + \|p\|_{\tilde{S}_0^{-1}}^2.$$

To prove the lower bound we observe that by the triangle inequality $\|u\|_{\tilde{A}_0^{-1}} \le \|u - dB^T S_0^{-1} p\|_{\tilde{A}_0^{-1}} + \|B^T S_0^{-1} p\|_{\tilde{A}_0^{-1}}$ and by Lemma 3.1

$$\|u\|_{\tilde{A}_0^{-1}}^2 + \|p\|_{\tilde{S}_0^{-1}}^2 \le 2(1+b_0^2)(\|u - dB^T S_0^{-1} p\|_{\tilde{A}_0^{-1}}^2 + \|p\|_{\tilde{S}_0^{-1}}^2).$$

The upper bound follows similarly.   ∎

Proof. (Proof of Theorem 3.1, primal preconditioner case.) By our assumptions matrix Q_p is symmetric positive definite. The spectrum of $\mathcal{P}_p^{-1}\mathcal{M}$ is real, as this matrix is similar to a symmetric real matrix

$$Q_p^{1/2}\mathcal{P}_p^{-1}\mathcal{M}Q_p^{-1/2} = Q_p^{-1/2}\mathcal{M}\mathcal{P}_p^{-T}(\mathcal{H}_p\mathcal{P}_p^{-1}\mathcal{M})\mathcal{P}_p^{-1}\mathcal{M}Q_p^{-1/2}.$$

In order to estimate the absolute value of the eigenvalues, we shall bound their squares by showing that

$$\frac{m_0^2}{4(1+b_0^2)^2}x^T \mathcal{J}x \le x^T \mathcal{M}^T(\mathcal{P}_p^{-1})^T \mathcal{J}\mathcal{P}_p^{-1}\mathcal{M}x \le 4m_1^2(1+b_0^2)^2 x^T \mathcal{J}x \tag{3.8}$$

for all $x \in R^n \times R^m$. In order to prove the upper bound, we apply Lemmas 3.2 and 3.3 to obtain

$$\begin{aligned} x^T \mathcal{M}^T(\mathcal{P}_p^{-1})^T \mathcal{J}\mathcal{P}_p^{-1}\mathcal{M}x &\le 2(1+b_0^2)x^T \mathcal{M}^T \mathcal{U}_p^T I^T \mathcal{J}^{-1} I \mathcal{U}_p \mathcal{M}x \\ &\le 4(1+b_0^2)^2 x^T \mathcal{M}^T \mathcal{J}^{-1}\mathcal{M}x \\ &\le 4m_1^2(1+b_0^2)^2 x^T \mathcal{J}x, \end{aligned}$$

where the last inequality is a consequence of (3.1). The lower bound follows in a similar way.   ∎

3.3   PCR convergence

Let us recall the general convergence result for the PCR method.

Proposition 3.3 (see [46]). *Let x^* denote the solution to* (1.1) *and let x_k be the k-th iteration of the PCR method* Odir$(Q, \mathcal{H}^{-1}, \mathcal{H}\mathcal{P}^{-1}\mathcal{M})$ *(see Algorithm 1) with Q defined in Corollary 3.1. Then*

$$\|x^* - x_k\|_Q \le 2\left(\frac{\kappa - 1}{\kappa + 1}\right)^{k/2}\|x^* - x_0\|_Q, \tag{3.9}$$

where κ is the spectral condition number of $\mathcal{P}^{-1}\mathcal{M}$,

$$\kappa = \frac{\max\{|\lambda| : \lambda \in \sigma(\mathcal{P}^{-1}\mathcal{M})\}}{\min\{|\lambda| : \lambda \in \sigma(\mathcal{P}^{-1}\mathcal{M})\}}.$$

Theorem 3.1 immediately yields the convergence result for the PCR preconditioned with either $\mathcal{P}_d$ or $\mathcal{P}_p$:

Corollary 3.2. *Under assumptions of Section 3.1, the spectral condition number κ of either $\mathcal{P}_d^{-1}\mathcal{M}$ or $\mathcal{P}_p^{-1}\mathcal{M}$ is bounded by*

$$\kappa \leq 4\frac{m_1}{m_0}(1+b_0^2)^2.$$

Proposition 3.3 refers to norm $\|\cdot\|_Q$ which, in view of its definition in Corollary 3.1, may look quite exotic. However, both Q_d and Q_p are spectrally equivalent to $\mathcal{J}$, as stated in the following theorem:

Theorem 3.2. *Let Q denote either Q_d in the case of dual block preconditioner, or Q_p in the primal case, cf. Corollary 3.1. Under assumptions of Section 3.1, Q is spectrally equivalent to $\mathcal{J}$:*

$$\frac{m_0}{2(1+b_0^2)h_1}\|x\|_J \leq \|x\|_Q \leq \frac{2m_1(1+b_0^2)}{h_0}\|x\|_J \qquad \forall x \in R^n \times R^m. \tag{3.10}$$

Proof. From the definition, $\|x\|_Q^2 = \|\mathcal{P}^{-1}\mathcal{M}x\|_{\mathcal{H}}^2$. By assumption (3.3) together with (3.8) it follows that

$$\|\mathcal{P}^{-1}\mathcal{M}x\|_{\mathcal{H}} \leq \frac{1}{h_0}\|\mathcal{P}^{-1}\mathcal{M}x\|_J \leq \frac{2m_1(1+b_0^2)}{h_0}\|x\|_J,$$

which proves the upper bound. The proof of the lower bound is analogous. ∎

4 Examples of $\mathcal{P}_d$ and $\mathcal{P}_p$

The family of preconditioners introduced in (1.4) and (1.5) includes several types of block preconditioners previously introduced in the literature. Table **1** presents a list of some such preconditioners, accompanied with the appropriate choice of parameters used in this chapter. It turns out that many of them can be applied also in the case when A is indefinite and $C = 0$, i.e. to generalized saddle point problems.

Here, and in what follows, if two matrices X, Y satisfy $x^T X x \geq x^T Y x$ for all x, we write for short $X \geq Y$.

Proposition 3.4. *Let $C = 0$, $\varepsilon_1 = 1$ and assume (3.1) and (3.2). Then Theorems 3.1 and 3.2 hold for preconditioners listed in Table **1** under additional conditions listed in Table **2**. The constant h_1 which appears in Theorem 3.2 will depend on α and β in Table **2**.*

Proof. Let $\mathcal{H}$ be given either by (2.3) in the dual case or by (2.4) in the primal case. Proposition 3.1 yields the symmetry of $\mathcal{H}$, therefore we only need to prove (3.3). Since the lower bound is already guaranteed by Proposition 3.2, it remains to check the upper bound in (3.3) and this is where we will need assumptions from Table **2**.

For the primal based penalty preconditioner, $\delta = -1$ and

$$\mathcal{H}_p = \begin{pmatrix} A - \tilde{A}_0 + B^T \tilde{S}_0^{-1} B & \\ & \tilde{S}_0 - C \end{pmatrix}.$$

---

[2]May also be treated as a special case of $\mathcal{P}_p$.
[3]With $\tilde{S}_0 = S_0 = B\tilde{A}_0^{-1}B^T$.

Type	Form of $\mathcal{P}_d^2$	c	d	ε_2
diagonal [5, 10, 12]	$\begin{pmatrix} \tilde{A}_0 & \\ & \tilde{S}_0 \end{pmatrix}$	0	0	1
upper triangular positive [17, 47]	$\begin{pmatrix} \tilde{A}_0 & B^T \\ & \tilde{S}_0 \end{pmatrix}$	0	1	1
lower triangular [18]	$\begin{pmatrix} \tilde{A}_0 & \\ B & -\tilde{S}_0 \end{pmatrix}$	1	0	-1

Type	Form of $\mathcal{P}_d$	c	d	ε_2
symmetric indefinite [8, 20, 21]	$\begin{pmatrix} \tilde{A}_0 & B^T \\ B & B\tilde{A}_0^{-1}B^T - \tilde{S}_0 \end{pmatrix}$	1	1	-1
doubly constrained[3] [48]	$\begin{pmatrix} \tilde{A}_0 & B^T \\ B & 2B\tilde{A}_0^{-1}B^T \end{pmatrix}$	1	1	1

Type	Form of $\mathcal{P}_p$	c	d	ε_2
primal-based penalty [9]	$\begin{pmatrix} \tilde{A}_0 - B^T\tilde{S}_0^{-1}B & B^T \\ B & -\tilde{S}_0 \end{pmatrix}$	1	1	-1

Table **1**: Various block preconditioners encountered in the literature and their relation to parameters in (1.4) and (1.5). In all cases, $\varepsilon_1 = 1$.

(Let us notice that this is exactly the inner product matrix used in [9]). Since $\tilde{S}_0$ is positive definite and we assumed $A \geq \alpha\tilde{A}_0$ for $\alpha > 1$, cf. [9, Theorem 3.3], then

$$A - \tilde{A}_0 + B^T\tilde{S}_0^{-1}B \geq (\alpha - 1)\tilde{A}_0$$

so that

$$\min\{(\alpha - 1), 1\}\, x^T \mathcal{J}x \leq x^T \mathcal{H}_p x \qquad \forall x \in R^n \times R^m.$$

Other cases can be treated in an analogous way; we refer the reader to [33] for details.   ∎

4.1   Pros and cons of scaling

It should be noticed that for certain block preconditioners one can provide conditions which actually lead to a system with positive spectrum, which in turn can usually be solved faster [33] and with the conjugate gradient method.

For example, from [18] it follows that the assumption on the lower triangular preconditioner in Table **2** already ensures that $\mathcal{H}_d\mathcal{P}_d^{-1}\mathcal{M}$ is positive definite. Similarly, for the symmetric indefinite preconditioner $c = d = 1$ analysed in [20, Theorems 5.1 (for $\delta = -1$) and 5.2 (for $\delta = 1$)] in the case of positive semidefinite A it has been proved that the above conditions are also sufficient for the preconditioned system $\mathcal{H}_d\mathcal{P}_d^{-1}\mathcal{M}$ to be positive definite, provided $A > 0$. The primal based penalty preconditioner has also been shown to lead to a positive definite system $\mathcal{H}_p\mathcal{P}_p^{-1}\mathcal{M}$ under certain additional assumptions that require proper scaling of A_0 and S_0, see [9, Theorem 3.3] for details.

The scaling conditions on A_0 and S_0 require a preprocessing step, in which one finds scaling factors which make the assumptions such as those in Table **2** satisfied. Usually this can be done by running a few iterations of the power method to estimate spectrum bounds, but sometimes the conditions are a priori satisfied [9]. It is not very much clear, however, how to choose "best" scaling parameters in

Type	$\tilde{A}_0$	$\tilde{S}_0$	δ
diagonal	none		1
upper triangular positive	none		1
lower triangular	$\exists \alpha > 1 \ A \geq \alpha \tilde{A}_0$	none	-1
symmetric indefinite	$\exists \alpha < 1 \ \alpha \tilde{A}_0 \geq A$	$\exists \beta > 1 \ B\tilde{A}_0^{-1}B^T \geq \beta \tilde{S}_0$	1
symmetric indefinite	$\exists \alpha > 1 \ A \geq \alpha \tilde{A}_0$	$\exists \beta < 1 \ \beta \tilde{S}_0 \geq B\tilde{A}_0^{-1}B^T$	-1
doubly constrained	$\exists \alpha < 1 \ \alpha \tilde{A}_0 \geq A$	$\tilde{S}_0 = B\tilde{A}_0^{-1}B^T$	1
primal based penalty	$\exists \alpha > 1 \ A \geq \alpha \tilde{A}_0$	none	-1

Table **2**: Additional assumptions in the case when $C = 0$, see Proposition 3.4 for details. Preconditioner types defined in Table **1**.

practical applications, as they not only alter the convergence speed, but also the norm in which the PCR minimizes the residual. It is thus interesting to note that Theorem 3.3 provides a selection of preconditioners and their defining parameters c, d and $\delta, \varepsilon_1, \varepsilon_2$ such that no additional scaling of A_0 or S_0 is necessary, making the PCR a truly parameter-free method.

Theorem 3.3. *Let assumptions* (3.1) *and* (3.2) *be satisfied with* $\varepsilon_1 = \varepsilon_2 = \delta = 1$. *Then Theorems 3.1 and 3.2 hold for preconditioners listed in Table* **3** *together with additional restrictions on A or C.*

Proof. From our assumptions and Proposition 3.2 it suffices to prove the upper bound in (3.3) only. For block diagonal or triangular preconditioners we have

$$\mathcal{H}_d = \begin{pmatrix} A_0 - cA & \\ & S_0 - dC \end{pmatrix}.$$

For the primal based penalty negative preconditioner, $c = d = -1$, so that

$$\mathcal{H}_p = \begin{pmatrix} A_0 + B^T \tilde{S}_0^{-1} B + A & \\ & S_0 - C \end{pmatrix},$$

while for the symmetric indefinite negative preconditioner we again have $c = d = -1$, but

$$\mathcal{H}_d = \begin{pmatrix} A_0 + A & \\ & S_0 + B\tilde{A}_0^{-1}B^T - C \end{pmatrix}.$$

In any case, the result is straightforward. ∎

5   Example Applications to PDEs

In this section we show three example applications of the theory developed above to some variational problems arising from the finite element discretization of partial differential equations. Following [2], we shall use a formulation which reflects the dependence of (1.1) on the mesh parameter h [5, 21, 30, 33].

Our goal is to provide a set of conditions under which the convergence rate of the PCR method for such problems can be bounded independently of the mesh size h. It will turn out that the key ingredients to construct a good preconditioner would be $\tilde{A}_0$ and $\tilde{S}_0$ which define a norm $\|\cdot\|_{\mathcal{J}}$, equivalent (uniformly in h) to the norm in which the system is stable and continuous (uniformly in h). Usually the task of finding a preconditioner to operators which define such a norm is much easier and in many cases such $\tilde{A}_0$ and $\tilde{S}_0$ are already known, analyzed and implemented in software packages.

Type	Form of $\mathcal{P}$	Additional restrictions on A, C
diagonal	$\begin{pmatrix} \tilde{A}_0 & \\ & \tilde{S}_0 \end{pmatrix}$	none
upper triangular positive	$\begin{pmatrix} \tilde{A}_0 & B^T \\ & \tilde{S}_0 \end{pmatrix}$	none
upper triangular negative	$\begin{pmatrix} \tilde{A}_0 & -B^T \\ & \tilde{S}_0 \end{pmatrix}$	$C \geq 0$
lower triangular negative	$\begin{pmatrix} \tilde{A}_0 & \\ -B & \tilde{S}_0 \end{pmatrix}$	$A \geq 0$
symmetric indefinite negative	$\begin{pmatrix} \tilde{A}_0 & -B^T \\ -B & \tilde{S}_0 + B\tilde{A}_0^{-1}B^T \end{pmatrix}$	$A \geq 0$ and $C \leq 0$
primal based penalty negative	$\begin{pmatrix} \tilde{A}_0 + B^T\tilde{S}_0 B & -B^T \\ -B & \tilde{S}_0 \end{pmatrix}$	$A \geq 0$ and $C \leq 0$

Table **3**: List of some preconditioners which do not require scaling of A_0 or S_0.

5.1   Generalized saddle point problems

(a) Framework and assumptions. Let V, W be real infinite dimensional Hilbert spaces with scalar products denoted by $((\cdot, \cdot))$ and $(\cdot, \cdot)$, respectively. The norms in these spaces, induced by the inner products, will be denoted by $||\cdot||$ and $|\cdot|$. The dual pairing between a Hilbert space and its dual will be denoted $\langle\langle \cdot, \cdot \rangle\rangle$.

We consider a family of finite dimensional subspaces indexed by the parameter $h \in (0,1)$: $V_h \subset V$, $\quad W_h \subset W$. If V_h, W_h come from a finite element approximation, the dimension of these subspaces increases for decreasing h.

Let us introduce two continuous bilinear forms: $a : V \times V \to R$, $b : V \times W \to R$, with norms $||a||_{V \times V}$, $||b||_{V \times W}$. We assume that $a(\cdot, \cdot)$ is symmetric and there exists a constant α_0, independent of h, such that

$$\exists \alpha_0 > 0 \quad \forall h \in (0,1) \quad \inf_{v \in V_h^0, v \neq 0} \sup_{u \in V_h^0, u \neq 0} \frac{a(u,v)}{||u|| \, ||v||} \geq \alpha_0, \tag{5.1}$$

where $V_h^0 = \{v \in V_h : \forall q \in W_h \quad b(v,q) = 0\}$. We shall also assume that the finite dimensional spaces V_h and W_h satisfy the uniform inf-sup condition,

$$\exists \beta_0 > 0 \quad \forall h \in (0,1) \quad \forall p \in W_h \quad \sup_{v \in V_h, v \neq 0} \frac{b(v,p)}{||v||} \geq \beta_0 |p|. \tag{5.2}$$

The variational finite-dimensional problem reads:

Problem 5.1. Find $(u_h, p_h) \in V_h \times W_h$ such that

$$\begin{cases} a(u_h, v_h) + b(v_h, p_h) = \langle\langle f, v_h \rangle\rangle, \\ b(u_h, q_h) = \langle\langle g, q_h \rangle\rangle, \end{cases}$$

for all $(v_h, q_h) \in V_h \times W_h$.

Introducing bases, $V_h = \text{span}\{\phi_1, \ldots, \phi_n\}$ and $W_h = \text{span}\{\eta_1, \ldots, \eta_m\}$, and forming the Gramian matrices for the corresponding bilinear forms, $A_{ij} = a(\phi_j, \phi_i)$, $B_{kj} = b(\phi_j, \eta_k)$, and the right hand side components $F_j = \langle\langle f, \phi_j \rangle\rangle$, $G_l = \langle\langle g, \eta_l \rangle\rangle$, we end up with a system of equations of the form (1.1):

Problem 5.2. Find $(u, p) \in R^n \times R^m$ such that

$$\mathcal{M}\begin{pmatrix} u \\ p \end{pmatrix} \equiv \begin{pmatrix} A & B^T \\ B & 0 \end{pmatrix}\begin{pmatrix} u \\ p \end{pmatrix} = \begin{pmatrix} F \\ G \end{pmatrix}. \tag{5.3}$$

To distinguish between vectors in $V_h \times W_h$ and their basis representations in $R^n \times R^m$, we drop the subscript h while referring to the latter.

(b) A family of block preconditioners for Problem 5.2 with convergence rate bounded uniformly in h.
Using the results of Section 3, we arrive at the following general condition which guarantees that a PCR method with preconditioners $\mathcal{P}_d$ or $\mathcal{P}_p$, as defined in Section 2, will be convergent with a rate bounded independently of h (see also [33]).

To simplify the notation, we shall use the following symbols. For x, y, the symbol $x \lesssim y$ shall mean that there exists a positive constant Const, independent of x, y, the mesh size h, the parameters c and d in (1.4) and δ in (2.3), such that $x \leq \text{Const } y$. (This constant is allowed to depend on other parameters of the problem, such as the continuity constants, etc.) Similarly, we write $x \gtrsim y$ if and only if $y \lesssim x$. Finally, $x \simeq y$ will denote that both $x \lesssim y$ and $x \gtrsim y$ hold.

Theorem 3.4. *Let us assume that Problem 5.1 satisfies the assumptions listed in paragraph (a), and suppose that*

$$\tilde{A}_0 \simeq L \quad \text{and} \quad \tilde{S}_0 \simeq M,$$

where L and M are induced by the norms in V_h, W_h respectively, so that $L_{ij} = ((\phi_j, \phi_i))$ and $M_{kl} = (\eta_l, \eta_k)$. Let us further assume that (3.3) holds with constants independent of h. Then the convergence rate of the PCR (see Algorithm 1), using either $\mathcal{P}_d$ or $\mathcal{P}_p$ as the preconditioner, is bounded independently of h.

Proof. It is sufficient to show that the assumptions of Theorem 3.1 are satisfied with constants independent of h. Using the correspondence between $(u_h, p_h) \in V_h \times W_h$ and its basis representation $(u, p) \in R^n \times R^m$, the following well known result [2] on the stability and continuity of (5.3) holds for $G \in \text{Im} B$:

$$\|u\|_L + \|p\|_M \simeq \|F\|_{L^{-1}} + \|G\|_{M^{-1}}. \tag{5.4}$$

From the continuity assumption on $b(\cdot, \cdot)$ it follows

$$p^T Bu = b(u_h, p_h) \leq \|b\|_{V \times W}\|u_h\|\,\|p_h\| = \|b\|_{V \times W}\|u\|_L\|p\|_M \lesssim \|u\|_{\tilde{A}_0}\|p\|_{\tilde{S}_0},$$

so that b_0 in (3.2) is independent of h. Moreover, from the spectral equivalence $\tilde{A}_0 \simeq L$ and $\tilde{S}_0 \simeq M$ and (5.4) we conclude that assumption (3.1) is satisfied with constants independent of h. ∎

(c) Application: the Stokes equation. Application of Theorem 3.4 to optimization problems has been discussed in [21, 33]. Other applications include problems where the (1,1) matrix features a large

kernel [49]. Here we will show how Theorem 3.4 applies to a simple model problem: a stable[4] discretization of Stokes equation [1] in a polygon $\Omega \subset R^2$:

$$\begin{cases} -\Delta u + \nabla p = f, \\ \text{div } u = 0, \end{cases} \qquad (5.5)$$

with homogeneous Dirichlet boundary conditions on u. The weak formulation is to find $(u, p) \in V \times W \equiv [H_0^1(\Omega)]^2 \times L_0^2(\Omega)$ such that

$$\begin{cases} (\nabla u, \nabla v)_{L^2(\Omega)} - (\text{div } v, p)_{L^2(\Omega)} = (f, v) & \forall v \in [H_0^1(\Omega)]^2, \\ (\text{div } u, q)_{L^2(\Omega)} = 0 & \forall q \in L_0^2(\Omega) \end{cases} \qquad (5.6)$$

for $f \in L^2(\Omega)$. Here and in the remaining part of this section, we use the notation $(\cdot, \cdot)_{L^2(\Omega)}$ to denote the inner product in $L^2(\Omega)$. The bilinear forms: $a(u, v) = (\nabla u, \nabla v)_{L^2(\Omega)}$ and $b(v, q) = -(\text{div } v, q)_{L^2(\Omega)}$ are continuous in the natural norms in V, W. We discretize (5.6) on a shape regular grid with diameter h, using for V_h, W_h any inf-sup stable pair of finite element spaces [1], so that condition (5.2) is satisfied.

In view of Theorem 3.4, the linear system resulting from the above discretization can be preconditioned by $\mathcal{P}_d$ or $\mathcal{P}_p$ with

- $\tilde{A}_0$ — spectrally equivalent to the discretized Laplace operator L with homogeneous Dirichlet boundary conditions; for example, a very efficient $\tilde{A}_0$ can be constructed using multigrid techniques [50] or domain decomposition methods [44].

- $\tilde{S}_0$ — spectrally equivalent to the mass matrix M, for example, a (diagonal) lumped mass matrix [51].

If assumption (3.3) is satisfied uniformly in h, then the convergence rate of the PCR is bounded independently of h. According to Proposition 3.4 and Theorem 3.3, in order to fulfill (3.3) for certain combinations of $c, d, \varepsilon_1, \varepsilon_2, \delta$, the matrices $\tilde{A}_0$ or $\tilde{S}_0$ may have to be adequately scaled.

5.2 Penalized saddle point problems

(a) Framework and assumptions. Within the framework of Section 5.1 paragraph (a), we consider a variational problem in $V_h \times W_h \subset V \times W$:

Problem 5.3. Find a pair $(u_h, p_h) \in V_h \times W_h$ which satisfies

$$\begin{cases} a(u_h, v_h) + b(v_h, p_h) = \langle\langle f, v_h \rangle\rangle, \\ b(u_h, q_h) - \varepsilon c(p_h, q_h) = \langle\langle g, q_h \rangle\rangle, \end{cases}$$

for all $(v_h, q_h) \in V_h \times W_h$.

Here, $\varepsilon \in [0, 1]$ is a given small parameter, sometimes called the penalty parameter. In this formulation, we still assume that all forms: $a(\cdot, \cdot), b(\cdot, \cdot), c(\cdot, \cdot)$ are continuous in the relevant norms in V, W and that both $a(\cdot, \cdot)$ and $c(\cdot, \cdot)$ are symmetric. We replace (5.1) with a stronger assumption that $a(\cdot, \cdot)$ is positive semidefinite on V_h, $a(u_h, u_h) \geq 0$ for all $u \in V_h$, and uniformly definite on V_h^0, i.e.

$$\exists a_0 > 0 \quad \forall h \in (0, 1) \quad \inf_{u_h \in V_h^0, u_h \neq 0} \frac{a(u_h, u_h)}{\|u_h\|^2} \geq a_0. \qquad (5.7)$$

---

[4]See Section 5.3 for an example of a *stablized* discretization.

In addition to the uniform inf-sup condition (5.2), we assume that $c(\cdot, \cdot)$ is positive semidefinite on W_h:

$$\forall p_h \in W_h \qquad c(p_h, p_h) \geq 0. \tag{5.8}$$

It is implicitly assumed that all mentioned above constants do not depend on ε.

(b) A family of block preconditioners for Problem 5.3 with convergence rate bounded uniformly in h and ε. Using the results of Section 3 we arrive at the following general condition which guarantees that a PCR method with preconditioners $\mathcal{P}_d$ or $\mathcal{P}_p$ will be convergent with a rate bounded independently of both h and ε.

Theorem 3.5. *Let us assume that Problem 5.3 satisfies the assumptions listed in paragraph (a) and suppose that* $\tilde{A}_0 \simeq L$ *and* $\tilde{S}_0 \simeq M$, *with L and M induced by the norms in* V_h, W_h, *so that* $L_{ij} = ((\phi_j, \phi_i))$ *and* $M_{kl} = (\eta_l, \eta_k)$. *Let us further assume that (3.3) holds with constants independent of h and* ε. *Then the convergence rate of the PCR (see Algorithm 1) using either* $\mathcal{P}_d$ *or* $\mathcal{P}_p$ *as the preconditioner is bounded independently of h and* ε.

Proof. In view of Proposition 3.2 and the proof of Theorem 3.4, it only remains to prove the stability of the solutions with a constant bounded uniformly in both h and ε. This is exactly the result that can be found e.g. in [4, Theorem III.4.11]. ∎

(c) Application to linear elasticity. The above result, which for block diagonal preconditioners probably has first been proved by Klawonn [5], can be applied e.g. to mixed formulations of linear elasticity equations, with the penalty parameter ε related to Lamé constants. See [5,9] for details.

5.3 Stabilized Stokes equations

Sometimes the spaces V_h, W_h are chosen in such a way that the inf-sup condition (5.2) is not satisfied and then one can resort to a so-called stabilized method. As a model example let us consider a stabilized $P_1 - P_1$ discretization of Stokes equations (5.5) with continuous pressure approximation. Let $\mathcal{T}_h$ denote a shape-regular, quasi-uniform triangulation of a polygonal $\Omega \subset R^2$ into triangles. Define the finite dimensional spaces of linear finite elements:

$$V_h = \{v \in [H_0^1(\Omega)]^2 : v_{|\kappa} \in [P_1(\kappa)]^2 \quad \forall \kappa \in \mathcal{T}_h\}$$

and

$$W_h = \{q \in L_0^2(\Omega) \cap C(\Omega) : q_{|\kappa} \in P_1(\kappa) \quad \forall \kappa \in \mathcal{T}_h\},$$

where $P_1(\kappa)$ denotes the space of linear functions on κ. Since V_h and W_h do not satisfy the inf-sup condition [2], let us consider the following stabilized discretization of (5.5), cf. [52]:

$$\begin{cases} (\nabla u_h, \nabla v_h)_{L^2(\Omega)} - (\operatorname{div} v_h, p_h)_{L^2(\Omega)} = (f, v_h)_{L^2(\Omega)} & \forall v_h \in V_h, \\ -(\operatorname{div} u_h, q_h)_{L^2(\Omega)} - c(p_h, q_h) = -\tau \sum_{\kappa \in \mathcal{T}_h} h_\kappa^2 (f, \nabla q_h)_{L^2(\kappa)} & \forall q_h \in W_h, \end{cases} \tag{5.9}$$

where

$$c(p_h, q_h) = \tau \sum_{\kappa \in \mathcal{T}_h} h_\kappa^2 (\nabla p_h, \nabla q_h)_{L^2(\kappa)}$$

and $\tau > 0$ is some prescribed parameter, independent of h. Thus, (5.9) has the form

$$\begin{cases} a(u_h, v_h) + b(v_h, p_h) = \langle\langle f, v_h \rangle\rangle \\ b(u_h, q_h) - c(p_h, q_h) = \langle\langle g, q_h \rangle\rangle, \end{cases}$$

where $a(\cdot, \cdot)$ and $b(\cdot, \cdot)$ are identical to those already defined in paragraph (c) for the stable discretization of Stokes system. Since by the inverse inequality [53] $c(\cdot, \cdot)$ is continuous in $L^2(\Omega)$ norm,

$$c(p, q) \leq \tau \sum_{\kappa \in \mathcal{T}_h} h_{\kappa}^2 |\nabla p|_{L^2(\kappa)} |\nabla q|_{L^2(\kappa)}$$

$$\leq \tau \sum_{\kappa \in \mathcal{T}_h} |p|_{L^2(\kappa)} |q|_{L^2(\kappa)} \lesssim |p|_{L^2(\Omega)} |q|_{L^2(\Omega)},$$

from the above and Proposition 3.2 it already follows that

$$\|\mathcal{M}x\|_{\mathcal{J}^{-1}} \lesssim \|x\|_{\mathcal{J}} \qquad \forall x \in R^n \times R^m,$$

where

$$\left\| \begin{pmatrix} u \\ p \end{pmatrix} \right\|_{\mathcal{J}}^2 = \|u\|_L^2 + \|p\|_M^2,$$

with L and M defined as in Theorem 3.4. Thus it remains to prove the stability result, i.e. the lower bound in (3.1). According to [52, Theorem 3.1], for any $\tau > 0$ the discrete problem satisfies

$$\sup_{\substack{(v_h, p_h) \in V_h \times W_h, \\ (v_h, q_h) \neq (0,0)}} \frac{a(u_h, v_h) + b(v_h, p_h) + b(u_h, q_h) - c(p_h, q_h)}{\|v_h\|_{[H_0^1(\Omega)]^2} + \|q_h\|_{L^2(\Omega)}} \gtrsim \|u_h\|_{[H_0^1(\Omega)]^2} + \|p_h\|_{L^2(\Omega)} \qquad (5.10)$$

for all $(u_h, p_h) \in V_h \times W_h$, which is nothing but the stability estimate in the norm $\left\| \begin{pmatrix} u \\ p \end{pmatrix} \right\|_{\mathcal{J}}$. Indeed, using the correspondence between the discrete variational problem and the algebraic formulation, we have that $\|u_h\|_{[H_0^1(\Omega)]^2} + \|p_h\|_{L^2(\Omega)} = \|u\|_L + \|p\|_M$ and

$$a(u_h, v_h) + b(v_h, p_h) + b(u_h, q_h) - c(p_h, q_h) = \begin{pmatrix} v \\ q \end{pmatrix}^T \mathcal{M} \begin{pmatrix} u \\ p \end{pmatrix},$$

so from (5.10) we conclude that

$$\|x\|_{\mathcal{J}} \lesssim \sup_{\substack{y \in R^n \times R^m, \\ y \neq 0}} \frac{y^T \mathcal{M} x}{\|y\|_{\mathcal{J}}} \leq \|\mathcal{M}x\|_{\mathcal{J}^{-1}}$$

for all $x \in R^n \times R^m$. Therefore again, if $\tilde{A}_0 \simeq L$ are $\tilde{S}_0 \simeq M$ are such that (3.3) holds with constants independent of h, then the convergence rate of the PCR preconditioned with either $\mathcal{P}_d$ or $\mathcal{P}_p$ is bounded independently of h.

6   Conclusions

We have discussed two classes of block preconditioners for symmetric saddle point problems and provided eigenvalue estimates of the preconditioned system $\mathcal{P}^{-1}\mathcal{M}$ under a quite general assumption of the stability and continuity of the problem being solved. In the context of PDEs, basing upon this result, an iterative method optimal with respect to the mesh size h can be designed, which may reuse existing state-of-the-art preconditioners or fast solvers for certain elliptic problems. The key ingredients were the assumption that the stability and continuity constants of $\mathcal{M}$ are independent of h and that the preconditioning blocks $\tilde{A}_0$ and $\tilde{S}_0$ give rise to a norm in which the problem is stable and continuous uniformly in h. We have also distinguished some preconditioners who do not need scaling, an important feature for the robustness of the method.

While methods which *do* require scaling may converge faster, as they transform the system into a positive definite one, the optimal choice of scaling parameters remains an open question.

While the theory presented here is built around the optimality in h, it does not cover other important issues encountered in the solution of certain PDEs, e.g. the optimality of the preconditioner with respect to some parameter which enters a low order term in the $(1,1)$ block, see [13, 54, 55].

7   Acknowledgments

The author is indebted to the anonymous referee for remarks that helped to improve significantly the manuscript. The research has been financially supported in part by Polish Ministry of Science and Higher Education grant N N201 0069 33.

Bibliography

[1]   Girault V, Raviart PA. Finite element methods for Navier-Stokes equations. vol. 5 of Springer Series in Computational Mathematics. Berlin: Springer-Verlag; 1986.

[2]   Brezzi F, Fortin M. Mixed and hybrid finite element methods. vol. 15 of Springer Series in Computational Mathematics. New York: Springer-Verlag; 1991.

[3]   Greif C, Schötzau D. Preconditioners for the discretized time-harmonic Maxwell equations in mixed form. Numer Linear Algebra Appl 2007;14(4):281–297.

[4]   Braess D. Finite elements. Cambridge: Cambridge University Press; 1997. Theory, fast solvers, and applications in solid mechanics, Translated from the 1992 German original by Larry L. Schumaker.

[5]   Klawonn A. An optimal preconditioner for a class of saddle point problems with a penalty term. SIAM J Sci Comput 1998;19(2):540–552.

[6]   Smith BF, Bjørstad PE, Gropp WD. Domain decomposition. Cambridge: Cambridge University Press; 1996. Parallel multilevel methods for elliptic partial differential equations.

[7]   Benzi M, Golub GH, Liesen J. Numerical solution of saddle point problems. Acta Numer 2005;14:1–137.

[8]   Bank RE, Welfert BD, Yserentant H. A class of iterative methods for solving saddle point problems. Numer Math 1990;56(7):645–666.

[9]   Dohrmann CR, Lehoucq RB. A primal-based penalty preconditioner for elliptic saddle point systems. SIAM J Numer Anal 2006;44(1):270–282 (electronic).

[10] D'yakonov EG. Iterative methods with saddle operators. Dokl Akad Nauk SSSR 1987;292(5):1037–1041.

[11] Rusten T, Winther R. A preconditioned iterative method for saddlepoint problems. SIAM J Matrix Anal Appl 1992;13(3):887–904. Iterative methods in numerical linear algebra (Copper Mountain, CO, 1990).

[12] Silvester D, Wathen A. Fast iterative solution of stabilised Stokes systems. II. Using general block preconditioners. SIAM J Numer Anal 1994;31(5):1352–1367.

[13] Bramble JH, Pasciak JE. Iterative techniques for time dependent Stokes problems. Comput Math Appl 1997;33(1-2):13–30. Approximation theory and applications.

[14] Axelsson O, Neytcheva M. Eigenvalue estimates for preconditioned saddle point matrices. Numer Linear Algebra Appl 2006;13(4):339–360.

[15] Klawonn A. Block-triangular preconditioners for saddle point problems with a penalty term. SIAM J Sci Comput 1998;19(1):172–184. Special issue on iterative methods (Copper Mountain, CO, 1996).

[16] Simoncini V. Block triangular preconditioners for symmetric saddle-point problems. Appl Numer Math 2004;49(1):63–80.

[17] Cao ZH. Positive stable block triangular preconditioners for symmetric saddle point problems. Appl Numer Math 2007;57(8):899–910.

[18] Bramble JH, Pasciak JE. A preconditioning technique for indefinite systems resulting from mixed approximations of elliptic problems. Math Comp 1988;50(181):1–17.

[19] Vassilevski PS, Lazarov RD. Preconditioning mixed finite element saddle-point elliptic problems. Numer Linear Algebra Appl 1996;3(1):1–20.

[20] Zulehner W. Analysis of iterative methods for saddle point problems: a unified approach. Math Comp 2002;71(238):479–505 (electronic).

[21] Schöberl J, Zulehner W. Symmetric indefinite preconditioners for saddle point problems with applications to PDE-constrained optimization problems. SIAM J Matrix Anal Appl 2007;29(3):752–773 (electronic).

[22] Axelsson O. Preconditioning of indefinite problems by regularization. SIAM J Numer Anal 1979;16(1):58–69.

[23] Elman HC, Golub GH. Inexact and preconditioned Uzawa algorithms for saddle point problems. SIAM J Numer Anal 1994;31(6):1645–1661.

[24] Bramble JH, Pasciak JE, Vassilev AT. Analysis of the inexact Uzawa algorithm for saddle point problems. SIAM J Numer Anal 1997;34(3):1072–1092.

[25] Golub GH, Greif C. On solving block-structured indefinite linear systems. SIAM J Sci Comput 2003;24(6):2076–2092 (electronic).

[26] Golub GH, Greif C, Varah JM. An algebraic analysis of block diagonal preconditioner for saddle point systems. SIAM J Matrix Anal Appl 2005;27(3):779–792 (electronic).

[27] Benzi M, Liu J. Block preconditioning for saddle point systems with indefinite $(1,1)$ block. Int J Comput Math 2007;84(8):1117–1129.

[28] Cao ZH. Augmentation block preconditioners for saddle point-type matrices with singular $(1,1)$ blocks. Numer Linear Algebra Appl 2008;15(6):515–533.

[29] Ashby SF, Manteuffel TA, Saylor PE. A taxonomy for conjugate gradient methods. SIAM J Numer Anal 1990;27(6):1542–1568.

[30] Krzyzanowski P. On block preconditioners for nonsymmetric saddle point problems. SIAM J Sci Comput 2001;23(1):157–169 (electronic).

[31] Stoll M, Wathen A. Combination preconditioning and the Bramble-Pasciak^{+} preconditioner. SIAM J Matrix Anal Appl 2008;30(2):582–608.

[32] Dollar HS, Gould NIM, Stoll M, Wathen AJ. Preconditioning saddle-point systems with applications in optimization. SIAM J Sci Comput 2010;32(1):249–270.

[33] Krzyzanowski P. On block preconditioners for saddle point problems with singular or indefinite $(1,1)$ block. Numer Linear Algebra Appl 2011;18(1):123–140.

[34] Bramble JH. Multigrid methods. vol. 294 of Pitman Research Notes in Mathematics Series. Harlow: Longman Scientific & Technical; 1993.

[35] Hackbusch W. Multigrid methods and applications. vol. 4 of Springer Series in Computational Mathematics. Berlin: Springer-Verlag; 1985.

[36] Toselli A, Widlund O. Domain decomposition methods—algorithms and theory. vol. 34 of Springer Series in Computational Mathematics. Berlin: Springer-Verlag; 2005.

[37] Pavarino LF, Widlund OB. Balancing Neumann-Neumann methods for incompressible Stokes equations. Comm Pure Appl Math 2002;55(3):302–335.

[38] Li J, Widlund O. BDDC algorithms for incompressible Stokes equations. SIAM J Numer Anal 2006;44(6):2432–2455.

[39] Li J. A dual-primal FETI method for incompressible Stokes equations. Numer Math 2005;102(2):257–275.

[40] Braess D, Sarazin R. An efficient smoother for the Stokes problem. Appl Numer Math 1997;23(1):3–19. Multilevel methods (Oberwolfach, 1995).

[41] Verfürth R. A multilevel algorithm for mixed problems. SIAM J Numer Anal 1984;21(2):264–271.

[42] Wittum G. Multi-grid methods for Stokes and Navier-Stokes equations. Transforming smoothers: algorithms and numerical results. Numer Math 1989;54(5):543–563.

[43] Peisker P. A multilevel algorithm for the biharmonic problem. Numer Math 1985;46(4):623–634.

[44] Klawonn A, Pavarino LF. A comparison of overlapping Schwarz methods and block preconditioners for saddle point problems. Numer Linear Algebra Appl 2000;7(1):1–25.

[45] de Niet AC, Wubs FW. Two preconditioners for saddle point problems in fluid flows. Internat J Numer Methods Fluids 2007;54(4):355–377.

[46] Hackbusch W. Elliptic differential equations. vol. 18 of Springer Series in Computational Mathematics. Berlin: Springer-Verlag; 1992. Theory and numerical treatment, Translated from the author's revision of the 1986 German original by Regine Fadiman and Patrick D. F. Ion.

[47] Murphy MF, Golub GH, Wathen AJ. A note on preconditioning for indefinite linear systems. SIAM J Sci Comput 2000;21(6):1969–1972 (electronic).

[48] Forsgren A, Gill PE, Griffin JD. Iterative solution of augmented systems arising in interior methods. SIAM J Optim 2007;18(2):666–690.

[49] Greif C, Schötzau D. Preconditioners for saddle point linear systems with highly singular $(1, 1)$ blocks. Electron Trans Numer Anal 2006;22:114–121 (electronic).

[50] Elman HC. Multigrid and Krylov Subspace Methods for the Discrete Stokes Equations. Internat J Numer Methods Fluids. 1996;22(8):755–770.

[51] Wathen AJ. Realistic eigenvalue bounds for the Galerkin mass matrix. IMA J Numer Anal 1987;7(4):449–457.

[52] Franca LP, Hughes TJR, Stenberg R. Stabilised finite element methods for the Stokes problem. In: Nicolaides RA, Gunzburger MD, editors. Incompressible Computational Fluid Dynamics. Cambridge University Press; 1993. p. 87–108.

[53] Ciarlet PG. Basic error estimates for elliptic problems. In: Handbook of numerical analysis, Vol. II. Handb. Numer. Anal, II. Amsterdam: North-Holland; 1991. p. 17–351.

[54] Mardal KA, Winther R. Uniform preconditioners for the time dependent Stokes problem. Numer Math 2004;98(2):305–327.

[55] Olshanskii MA, Benzi M. An augmented Lagrangian approach to linearized problems in hydrodynamic stability. SIAM J Sci Comput 2008;30(3):1459–1473.

Chapter 4

Construction of Preconditioners by Mapping Properties for Systems of Partial Differential Equations

Kent-Andre Mardal[1] and Ragnar Winther[2]

Abstract: The purpose of this chapter is to discuss a general approach to the construction of preconditioners for the linear systems of algebraic equations arising from discretizations of systems of partial differential equations. The discussion here is closely tied to our earlier paper [1], where we gave a comprehensive review of a mathematical theory for constructing preconditioners based on the mapping properties of the coefficient operator of the underlying differential systems. In the presentation given below we focus more on specific examples, while just an outline of the general theory is given.

Keywords: *partial differential equations, discretizations, linear systems, algebraic, preconditioners, mathematical theory, Hilbert space, operator*

1   Introduction

Discretization methods for partial differential equations are often designed to mimic key properties of the problems they are approximating. For example, discretizations of conservation laws are frequently constructed such that corresponding discrete quantities are conserved, while finite element spaces typically inherit continuity properties from the requirement that they should be subspaces of the Sobolev spaces they approximate. In this chapter we will demonstrate that also iterative solution algorithms for discretized differential systems should inherit key properties from the corresponding continuous systems. In particular, we will argue that the mapping properties of the governing differential operators suggest the basic structure of efficient preconditioners for the corresponding discrete systems.

---

[1](corresponding author) Center for Biomedical Computing, Simula Research Laboratory, Lysaker, Norway; e-mail: kent-and@simula.no

[2]Centre of Mathematics for Applications and Department of Informatics, University of Oslo, Norway; e-mail: ragnar.winther@cma.uio.no

Owe Axelsson and János Karátson (Eds)

Linear differential operators usually have an unbounded spectrum, and as a consequence, standard iterative methods like the conjugate gradient and related Krylov space methods will not converge, or in many cases will not even be well defined, for such problems. These properties are reflected in the corresponding discrete problems. Even if these problems are of finite dimension, the spectrum of the coefficient operators will be unbounded as the mesh is refined, causing slower convergence of iterative methods for finer meshes. To overcome this effect, preconditioners are introduced. In fact, even the underlying differential equations can be transformed into problems which admit convergent iterations if the systems are properly preconditioned. As we will explain in Section 3 below, the natural preconditioner for such systems is an isomorphism mapping the space of right hand sides into the solution space. By applying the corresponding ideas to the discrete analogs we obtain so–called *canonical preconditioners* for the corresponding discrete systems. Typically, these preconditioners lead to bounded condition numbers, and therefore to rates of convergence which are bounded uniformly with respect to the discretization. However, these preconditioners are usually not computationally efficient, since they typically are composed of inverses of discrete differential operators. Computationally efficient preconditioners can often be constructed by replacing these inverses by corresponding analogs generated by techniques like multigrid or domain decomposition methods.

This chapter is closely tied to the review paper [1], where we present a comprehensive overview of an abstract theory to construct preconditioners for discrete systems from the mapping properties of the corresponding differential operators. This technique originates from the papers [2, 3] and has later been exploited in a variety of applications, see [1] and references given there. The strength of the approach to the construction of preconditioners for discretized differential systems presented here is most striking when we consider singular perturbation problems. For such problems we will construct preconditioners which behave uniformly well both with respect to the discretization and the perturbation parameter.

We remark that there are close similarities between the abstract approach to preconditioning taken here, and several other discussions of preconditioning. The relation between preconditioning of elliptic problems and the concept "equivalent operators" has been utilized by several authors, cf. for example [4, 5], while a more general approach to "operator preconditioning" is outlined in [6]. Furthermore, although the examples presented both in this chapter and in [1] are mostly symmetric saddle point problems, the theory is not limited to such applications. Some discussion of nonsymmetric problems are for example given in [7–9]. Alternative block preconditioners for saddle point problems, on triangular or indefinite form, are described in [10–18].

The numerical examples presented in this chapter are implemented in FEniCS [19] and are slight modifications of the examples described in [20]. In FEniCS we have used the linear algebra backend Trilinos [21] and its algebraic multigrid toolbox ML.

The outline of this chapter is as follows. In Section 2 we present a series of numerical examples based on discretizations of various differential systems, namely the Stokes problem, linear elasticity, the stabilized Stokes problem, and the time dependent Stokes problem. The general abstract framework given in [1] is briefly described in Section 3, before we revisit in Section 4 some of the examples studied previously.

2   Motivating Examples

In this section we will discuss a number of numerical experiments, where we present condition numbers and iteration counts for various discrete differential systems using different preconditioners. The purpose of these examples is to motivate the need for a better theoretical understanding on how to construct effective preconditioners. Such a theoretical overview will then be presented in the next section. Throughout the discussion below it is convenient to let the underlying system of partial differential equations be denoted $\mathcal{A}x = f$, where $\mathcal{A}$ is the governing differential operator, f is the given data, while x is the solution. Here, x and f are elements of appropriate function spaces. Furthermore, we will use $\mathcal{B}$ to denote various preconditioners at the continuous level.

In all the examples below the domain Ω is taken to be the unit square in $\mathbb{R}^2$, and we will use $\partial\Omega$ to denote the boundary of Ω. Most commonly the computations will be done with respect to a uniform triangular mesh, obtained by dividing Ω into $h \times h$ squares, where $h = 1/(N-1)$, and then dividing each square into two triangles. The corresponding discrete finite element systems are denoted $\mathcal{A}_h x_h = f_h$, and the corresponding preconditioners by $\mathcal{B}_h$. In the numerical examples below we will typically report estimates for the condition numbers of the operators $\mathcal{B}_h\mathcal{A}_h$. Furthermore, if $\mathcal{A}_h$ is positive definite we will give iteration counts for the conjugate gradient method (CG) applied to the preconditioned system $\mathcal{B}_h\mathcal{A}_h x_h = \mathcal{B}_h f_h$, while for indefinite problems we have used the conjugate gradient method on the normal equation (CGN). More precisely, we have used CG to solve

$$\mathcal{B}_h\mathcal{A}_h^T\mathcal{B}_h\mathcal{A}_h x = \mathcal{B}_h\mathcal{A}_h^T\mathcal{B}_h f_h.$$

This application of CGN requires that the preconditioner is symmetric and positive definite. Of course, since all our examples below are symmetric, we could also have used an iterative method which is more tailored, and usually more efficient, for such problems, cf. for example [22]. However, the purpose here is not efficiency, but the comparisons between different models and preconditioners. Therefore, we have preferred to use a robust method like CGN.

In some examples we compute the condition number by using a canonical preconditioner, i.e., the preconditioner is composed of inverses of discrete differential operators. However, since this approach requires the inversion of matrices, it is limited to coarse grids. On finer grids the canonical preconditioners are modified by introducing algebraic multigrid (AMG) operators as replacements for the exact inverses.

Example 4.1. *The Stokes problem.*
The Stokes problem for an incompressible fluid is:

$$-\Delta u - \operatorname{grad} p \;=\; f \quad \text{in } \Omega, \tag{2.1}$$
$$\operatorname{div} u \;=\; 0 \quad \text{in } \Omega, \tag{2.2}$$
$$u \;=\; 0 \quad \text{on } \partial\Omega, \tag{2.3}$$

where we refer to the unknown vector field u as the velocity and the unknown scalar field p as the pressure. We can write this system of equations formally as

$$\mathcal{A}\begin{pmatrix} u \\ p \end{pmatrix} = \begin{pmatrix} -\Delta & -\operatorname{grad} \\ \operatorname{div} & 0 \end{pmatrix}\begin{pmatrix} u \\ p \end{pmatrix} = \begin{pmatrix} f \\ 0 \end{pmatrix}.$$

The Stokes problem is a saddle point problem, and for a stable finite element discretization the proper inf–sup condition should be fulfilled. In our experiment we have used the lowest order Taylor–Hood element, i.e., continuous piecewise quadratic velocity fields, and continuous piecewise linear pressures. As a preconditioner we have used the discrete analog of the operator

$$\mathcal{B} = \begin{pmatrix} -\Delta^{-1} & 0 \\ 0 & I \end{pmatrix}$$

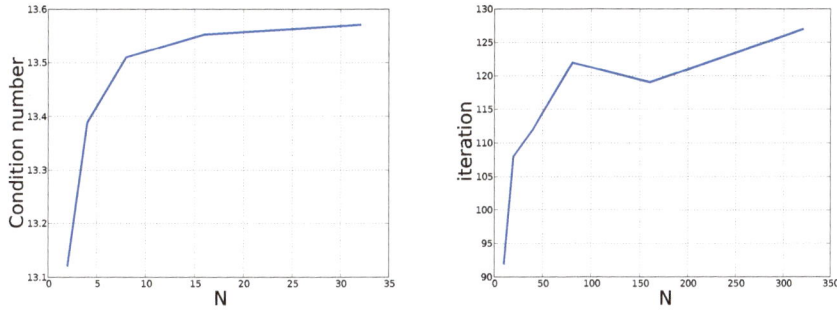

Figure **(1)**: The left figure shows the condition number of the preconditioned Stokes operator $\mathcal{B}_h\mathcal{A}_h$ for different meshes using the canonical precondititioners $\mathcal{B}_h$. The right figure shows the number of iterations required for convergence when using an AMG preconditioner combined with CGN for different meshes. The convergence criteria was a relative reduction of the preconditioned residual by a factor greater than 10^{10} in the discrete L_2 norm.

on these spaces. In Figure **(1)**, the condition number of the corresponding discrete analogs of the operator $\mathcal{B}\mathcal{A}$ and iterations counts for the preconditioned CGN are given. Clearly, the condition number and the number of iterations remains bounded as the mesh is refined. We remark that the condition number of the operator $\mathcal{A}_h$ in this example is $1.1 \cdot 10^6$ for $N = 32$, and that it grows like h^{-2}.

Example 4.2. *The linear elasticity problem.*
The elasticity problem for an isotropic material reads:

$$-(\lambda+\mu)\,\mathrm{grad}\,\mathrm{div}\,u - \mu\Delta u \;=\; f,\ \text{in}\ \Omega, \tag{2.4}$$
$$u \;=\; 0,\ \text{on}\ \partial\Omega, \tag{2.5}$$

where the positive constants λ and μ are the Lame's elasticity constants. For a nearly incompressible material $\lambda \gg \mu$, and for such problems the phenomena of *locking*, i.e., the case where the finite element computations produce significantly smaller deformation than it should, is well-known [23, 24]. Locking can be avoided by using reduced integration [25,26] or special finite element spaces [27]. Another consequence of $\lambda \gg \mu$ is that the condition numbers of the corresponding finite element matrices increase, and hence the convergence of iterative methods deteriorates. We first investigate the linear system obtained by a finite element approximation of the system (2.4)–(2.5) using continuous piecewise quadratics to approximate the displacement u. The parameter $\mu = 1$, while λ and the mesh parameter varies. In Figure **(2)** we show the number of iterations needed for convergence of CG combined with an AMG preconditioner. Clearly, the number of iterations increases as $\lambda \to \infty$.
One common way of avoiding locking is to introduce a separate variable for the divergence, i.e., we let $p = (\lambda+\mu)\,\mathrm{div}\,u$. For $\mu = 1$ the problem can then be written as

$$-\Delta u - \mathrm{grad}\,p \;=\; f,\ \text{in}\ \Omega,$$
$$\mathrm{div}\,u - \varepsilon^2 p \;=\; 0,\ \text{in}\ \Omega,$$
$$u \;=\; 0,\ \text{on}\ \partial\Omega,$$

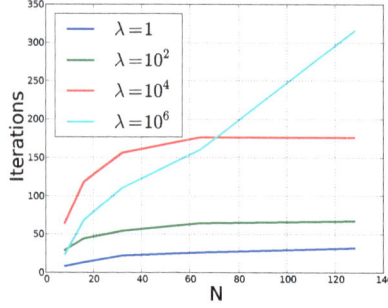

Figure **(2)**: The figure shows the number of iterations required for convergence with respect to different mesh resolutions and values of λ. We used CG combined with a standard AMG preconditioner. The convergence criteria was a relative reduction of the preconditioned residual of a factor greater than 10^6 in the preconditioned discrete L_2 norm.

where $\varepsilon^2 = 1/(1+\lambda)$. So as λ becomes large, ε tends to zero. This is a mixed system which formally can be written in the form

$$\mathcal{A}\begin{pmatrix} u \\ p \end{pmatrix} = \begin{pmatrix} -\Delta & -\operatorname{grad} \\ \operatorname{div} & -\varepsilon^2 I \end{pmatrix}\begin{pmatrix} u \\ p \end{pmatrix} = \begin{pmatrix} f \\ 0 \end{pmatrix}.$$

Hence, when λ is large, ε tends to zero and the problem formally approaches the Stokes system. Therefore, as in the previous example, we discretize the problem by the lowest order Taylor–Hood element. In particular, as above the velocity field is approximated by piecewise quadratics. We consider the same preconditioner as we used in Example 4.1, i.e., the discrete analog of

$$\mathcal{B}_1 = \begin{pmatrix} -\Delta^{-1} & 0 \\ 0 & I \end{pmatrix}.$$

Motivated by the diagonal elements in $\mathcal{A}$ we also consider the alternative preconditioner

$$\mathcal{B}_2 = \begin{pmatrix} -\Delta^{-1} & 0 \\ 0 & \varepsilon^{-2}I \end{pmatrix}.$$

These two preconditioners are compared with respect to different values of h and λ in Figure **(3)**. We note that the two preconditioners coincide when $\lambda = 1$. The leftmost graph in Figure **(3)** clearly demonstrates that the condition numbers of the discrete analogs of $\mathcal{B}_1\mathcal{A}$ are bounded independently of λ, while the condition numbers of the discrete analogs of $\mathcal{B}_2\mathcal{A}$ grows as λ increases. Similarly, in the rightmost graph the CGN combined with an AMG preconditioner of $\mathcal{B}_1$ gives convergence in a number of iterations that is bounded independently of both h and λ, while the preconditioner $\mathcal{B}_2$ behaves poorly.

Example 4.3. *The stabilized Stokes problem.*
Consider the Stokes problem from Example 4.1 once more. By perturbing the system slightly one can obtain stability of the discretization by using the same polynomial order of the finite element spaces for both velocity and pressure. One popular stabilization method consists of perturbing the equation for mass conservation by adding a small diffusion term on the pressure, i.e., we consider the system

$$\mathcal{A}_\varepsilon\begin{pmatrix} u \\ p \end{pmatrix} = \begin{pmatrix} -\Delta & -\operatorname{grad} \\ \operatorname{div} & \varepsilon^2\Delta \end{pmatrix}\begin{pmatrix} u \\ p \end{pmatrix} = \begin{pmatrix} f \\ g \end{pmatrix}. \tag{2.6}$$

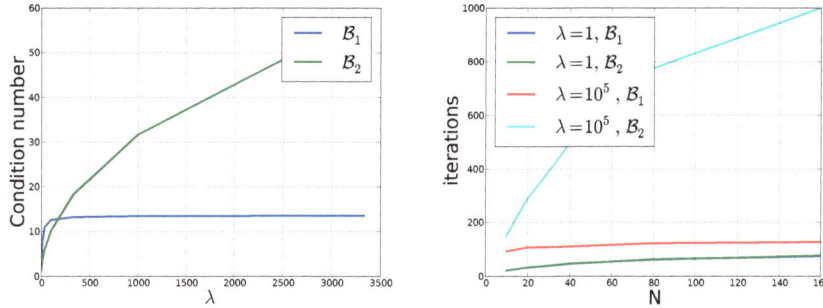

Figure (**3**): The left figure shows the condition number of the preconditioned discrete operators as a function of λ, using the two different canonical preconditititioners corresponding to $\mathcal{B}_1$ and $\mathcal{B}_2$. The right figure shows the number of iterations required for convergence with respect to the mesh resolutions, for two different values of λ, when using CGN with an AMG preconditioner. The convergence criteria was a relative reduction of the preconditioned residual of a factor greater than 10^{10} in the discrete L_2 norm.

It has been shown in e.g. [28, 29] that the choice $\varepsilon^2 = \beta h^2$, where $\beta > 0$ is fixed, gives a stable discretization with equal order elements in velocity and pressure. This method requires additional boundary conditions on the pressure which usually is assumed to be homogeneous Neumann conditions.

For this system we consider two preconditioners:

$$\mathcal{B}_1 = \begin{pmatrix} -\Delta^{-1} & 0 \\ 0 & I \end{pmatrix} \text{ and } \mathcal{B}_2 = \begin{pmatrix} -\Delta^{-1} & 0 \\ 0 & (-\varepsilon^2\Delta)^{-1} \end{pmatrix}.$$

Here, the $\mathcal{B}_1$ is the good Stokes preconditioner from above. This preconditioner is independent of ε, while the preconditioner $\mathcal{B}_2$ includes the diffusion term in the pressure. In the computations we have used piecewise linears for both velocity and pressure.

Figure (**4**) shows the behavior of the discrete versions of the preconditioners $\mathcal{B}_1$ and $\mathcal{B}_2$ in terms of h and β, where $\varepsilon^2 = \beta h^2$. Clearly, the ε–independent preconditioner $\mathcal{B}_1$, combined with a Krylov solver like CGN, produces an efficient algorithm that obtains convergence in a number of iterations that is bounded independently of h, for a given β. On the other hand, $\mathcal{B}_2$ shows poor performance.

Example 4.4. *The time dependent Stokes problem.*
The time dependent Stokes problem is an initial value problem of the form:

$$\begin{aligned}
u_t - \Delta u - \operatorname{grad} p &= f, \text{ in } \Omega, \, t > 0, \\
\operatorname{div} u &= 0, \text{ in } \Omega, \, t > 0, \\
u &= 0, \text{ on } \partial\Omega, \, t > 0, \\
u &= u_0, \text{ in } \Omega, \, t = 0.
\end{aligned}$$

Here, u is the unknown velocity vector, p is the uknown pressure, and u_0 is the initial condition. Various time stepping schemes applied to this equation result in systems of equations of the following form to be solved at each time step,

$$\mathcal{A}_\varepsilon \begin{pmatrix} u \\ p \end{pmatrix} = \begin{pmatrix} I - \varepsilon^2\Delta & -\operatorname{grad} \\ \operatorname{div} & 0 \end{pmatrix} \begin{pmatrix} u \\ p \end{pmatrix} = \begin{pmatrix} g \\ 0 \end{pmatrix},$$

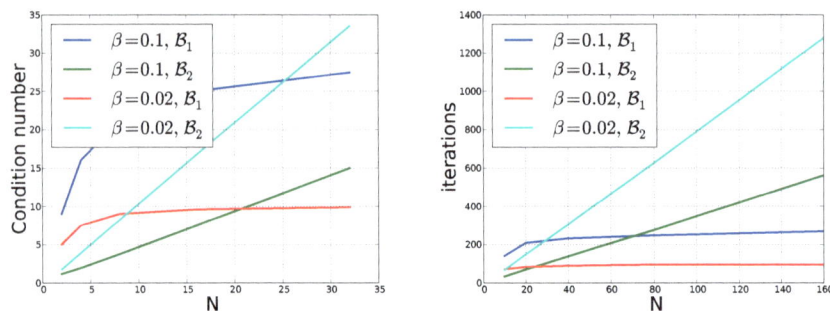

Figure **(4)**: The left figure shows the condition number of the preconditioned system using the two different canonical precondititioners $\mathcal{B}_1$ and $\mathcal{B}_2$ with respect to h, for two different values of β. The right figure shows the number of iterations required for convergence when using an AMG preconditioner combined with CGN for different meshes. The convergence criteria was a relative reduction of the preconditioned residual of a factor greater than 10^{10} in the discrete L_2 norm.

where g here includes both the right-hand side f and the solution at the previous time step. The small positive parameter ε is related to the time step.

Efficient preconditioners for this system have been described in [30–32]. These are of the form:

$$\mathcal{B}_\varepsilon = \begin{pmatrix} (I - \varepsilon^2 \Delta)^{-1} & 0 \\ 0 & (I - \Delta)^{-1} + \varepsilon^2 I \end{pmatrix}.$$

In this example, it is harder to "guess" the form of an efficient preconditioner by simple means. However, as we will see in the discussion below, this preconditioner arise naturally from the abstract reasoning in following section. Again the system is discretized by the lowest order Taylor–Hood method. In Figure **(5)** we present the condition numbers of the discrete preconditioned system and corresponding iteration counts. The observed condition numbers are bounded by 14, independently of ε and mesh refinements, and the number of iterations is bounded by 130. Similar results can be found in e.g. [1, 32].

3   Variational Problems and Preconditioning

In this section we will first briefly review the abstract approach to preconditioning outlined in [1], and thereafter we will discuss how this theory relates to the examples presented above. The main motivation for designing preconditioners for linear systems of equations is related to the use of iterative solution algorithms. Consider a linear system of the form

$$\mathcal{A}x = f, \tag{3.1}$$

where $\mathcal{A}$ is a bounded and invertible linear operator mapping a real, separable Hilbert space X into itself. In other words, we assume that $\mathcal{A}, \mathcal{A}^{-1} \in L(X, X)$, where in general $L(X, Y)$ is the set of bounded linear operators from X to Y. If the operator $\mathcal{A}$ is also symmetric and positive definite then the equation (3.1) can be solved by the CG iteration in the sense that the approximate solutions $\{x_m\}$ satisfy

$$\|x - x_m\|_{\mathcal{A}} \leq 2\alpha^m \|x - x_0\|_{\mathcal{A}}.$$

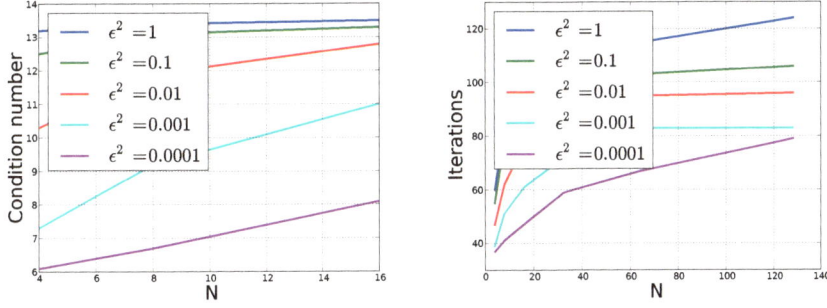

Figure (**5**): The left figure shows the condition number of the preconditioned system arising from discretizations of the time dependent Stokes problem, for different ε. The right figure shows the number of iterations required for convergence when using an AMG preconditioner combined with CGN for different mesh refinements and choices of ε. The convergence criteria was a relative reduction of the preconditioned residual of a factor greater than 10^{10} in the discrete L_2 norm.

Here, the convergence parameter α satisfies $0 \leq \alpha < 1$, $x_0 \in X$ is an abritary start vector, and the energy norm, $\|\cdot\|_{\mathcal{A}}$, is given by $\|x\|_{\mathcal{A}}^2 = \langle \mathcal{A}x, x \rangle$, where $\langle \cdot, \cdot \rangle$ denotes the inner product on X. More precisely, the convergence parameter α admits the bound

$$\alpha \leq \frac{\sqrt{\kappa(\mathcal{A})} - 1}{\sqrt{\kappa(\mathcal{A})} + 1},$$

where the condition number of the operator $\mathcal{A}$, $\kappa(\mathcal{A})$, is given as the product of the operator norms of $\mathcal{A}$ and $\mathcal{A}^{-1}$. In fact, x_m is the best approximation of the solution x in the associated Krylov space

$$K_m = K_m(\mathcal{A}, f) = \operatorname{span}\{f, \mathcal{A}f, \ldots, \mathcal{A}^{m-1}f\},$$

in the sense that

$$\|x - x_m\|_{\mathcal{A}} = \inf_{y \in K_m} \|x - y\|_{\mathcal{A}}.$$

A key to the efficiency of CG is that only one evaluation of the operator $\mathcal{A}$ is necessary to compute x_m from x_{m-1}.

If the operator $\mathcal{A}$ is a symmetric, but indefinite isomorphism on X, then we can instead apply CGN, i.e., CG to the normal equation $\mathcal{A}^2 x = \mathcal{A}b$. Also in this case the convergence rate depends on $\kappa(\mathcal{A})$, since $\kappa(\mathcal{A}^2) = \kappa(\mathcal{A})^2$ for symmetric operators. For more details on CG, CGN, and more general Krylov space methods we refer to [1, Section 2] and references given there.

3.1 Preconditioning

Krylov space methods like CG and CGN can in general not be applied directly to systems of partial differential equations, since the coefficient operators are not bounded. Consider for example the Stokes operator $\mathcal{A}$ studied in Example 2.1 above. This operator cannot be seen as a bounded operator of a Hilbert space into itself, since the eigenvalues accumulate at infinity. If the domain Ω is a bounded subset of $\mathbb{R}^n$, then the appropriate weak formulation of the operator $\mathcal{A}$ leads to a bounded operator from $X = (H_0^1(\Omega))^n \times L_0^2(\Omega)$ into $X^* = (H^{-1}(\Omega))^n \times L_0^2(\Omega)$. Here, $(H_0^1(\Omega))^n$ is the space of all L^2

vector fields with weak first order derivatives in L^2, and which are zero in the trace sense on the boundary $\partial\Omega$, while $L_0^2(\Omega)$ denote the set of L^2 scalar fields with mean value zero. Finally, the space $(H^{-1}(\Omega))^n \supset (L^2(\Omega))^n$ represents the dual of $(H_0^1(\Omega))^n$. In particular, $X \subsetneq X^*$. Since the operator $\mathcal{A}$ maps the solution space X out of itself, Krylov space methods like CG and CGN are in general not well–defined for such problems. However, if $\mathcal{B}$ is an operator such that $\mathcal{B} \in L(X^*, X)$ then the composition $\mathcal{B} \circ \mathcal{A} \in L(X, X)$, cf. Figure (**6**). Hence, we can apply Krylov space methods to the corresponding preconditioned equation

$$\mathcal{B}\mathcal{A}x = \mathcal{B}f.$$

Assume in general that $\mathcal{A} \in L(X, X^*)$ is an isomorhism, i.e., $\mathcal{A}^{-1} \in L(X^*, X)$. Here, X and X^* are assumed to be separable Hilbert spaces, where we should think of X^* as a representation of the dual of X. We also assume that $\mathcal{A}$ is symmetric in the sense that

$$\langle \mathcal{A}x, y \rangle = \langle x, \mathcal{A}y \rangle, \quad x, y \in X, \tag{3.2}$$

where $\langle \cdot, \cdot \rangle$ is the associated duality pairing between X and X^*.

We will assume that the preconditioner $\mathcal{B}$ is symmetric and positive definite in the sense that $\langle \cdot, \mathcal{B} \cdot \rangle$ is an inner product on X^*. Hence, the preconditioner is a *Riesz operator* mapping X^* to X. As a consequence, $\langle \mathcal{B}^{-1} \cdot, \cdot \rangle$ is an inner product on X. It is a direct consequence of these assumptions that the composition $\mathcal{B}\mathcal{A}$ is an isomorphism mapping X to itself. Furthermore, the operator $\mathcal{B}\mathcal{A} : X \to X$ is symmetric in the inner product $\langle \mathcal{B}^{-1} \cdot, \cdot \rangle$ on X. Therefore, the preconditioned system

$$\mathcal{B}\mathcal{A}x = \mathcal{B}f,$$

can be solved by CGN with a convergence rate bounded by $\kappa(\mathcal{B}\mathcal{A}) = \|\mathcal{B}\mathcal{A}\|_{L(X,X)} \|(\mathcal{B}\mathcal{A})^{-1}\|_{L(X,X)}$.

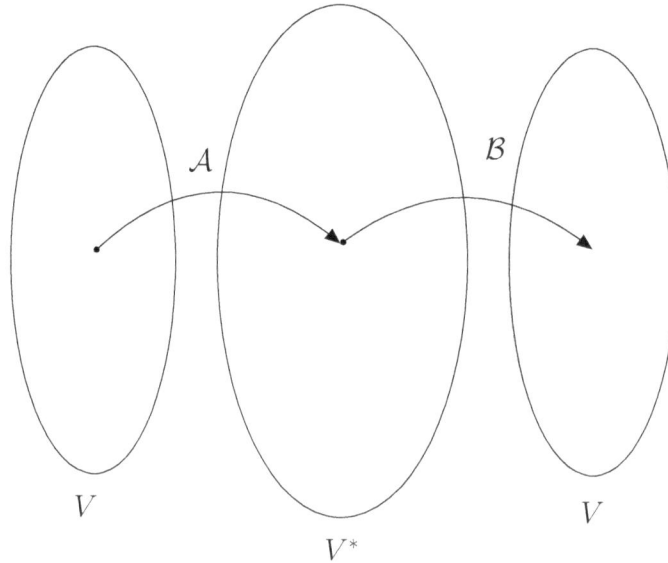

Figure (**6**): The mapping property of the composition of $\mathcal{A}$ and $\mathcal{B}$.

3.2 Variational problems

Many systems of partial differential equations can be formulated as variational problems. Throughout the discussion here, we will consider abstract variational problems of the form:
Find $x \in X$ such that for $f \in X^*$:

$$a(x,y) = \langle f, y \rangle, \quad y \in X, \tag{3.3}$$

where, as above, X is a Hilbert space, X^* its dual space, and $\langle \cdot, \cdot \rangle$ the associated duality pairing. Furthermore, we assume that $a : X \times X \to \mathbb{R}$ is a symmetric bilinear form. Following the classical variational theory of Babuška [33,34], the following two conditions, referred to as a boundedness condition and an inf–sup condition, are sufficient to guarantee existence, uniqueness, and well-posedness of the problem:

$$|a(x,y)| \leq c_1 \|x\|_X \|y\|_X, \quad x, y \in X. \tag{3.4}$$

and

$$\inf_{x \in X} \sup_{y \in X} \frac{a(x,y)}{\|x\|_X \|y\|_X} \geq c_2 > 0, \tag{3.5}$$

where c_1 and c_2 are positive constants.
We may write the problem (3.3) as a linear system of the form

$$\mathcal{A}x = f,$$

where $\mathcal{A} \in \mathcal{L}(X, X^*)$ is the linear operator defined by

$$\langle \mathcal{A}x, y \rangle = a(x,y), \quad x, y \in X.$$

Conditions (3.4) and (3.5) will imply that the operator $\mathcal{A}$ is an isomorphism mapping X to X^*. In particular,

$$\|\mathcal{A}\|_{\mathcal{L}(X,X^*)} \leq c_1 \text{ and } \|\mathcal{A}^{-1}\|_{\mathcal{L}(X^*,X)} \leq c_2^{-1}.$$

Furthermore, since the bilinear form a is symmetric, the operator $\mathcal{A}$ is symmetric in the sense of (3.2). The canonical preconditioner in this case is the Riesz operator mapping X^* to X, i.e., $\mathcal{B} \in \mathcal{L}(X^*, X)$ is defined by

$$\langle \mathcal{B}f, y \rangle_X = \langle f, y \rangle, \quad y \in X.$$

Here $\langle \cdot, \cdot \rangle_X$ denotes an inner product on the Hilbert space X. The operator $\mathcal{B}$ is symmetric and positive definite in the sense specified above. In fact, it is straightforward to verify that

$$\kappa(\mathcal{B}\mathcal{A}) = \|\mathcal{A}\|_{\mathcal{L}(X,X^*)} \cdot \|\mathcal{A}^{-1}\|_{\mathcal{L}(X^*,X)} \leq c_1/c_2, \tag{3.6}$$

and therefore the convergence rate of CGN, applied to the preconditioned system

$$\mathcal{B}\mathcal{A}x = \mathcal{B}f,$$

can be bounded in terms of the constants c_1 and c_2.
Next, we consider the discrete variational problems approximating the system (3.3). Hence, let $\{X_h\}$ be a family of subspaces of the Hilbert space X, where $h > 0$ is refered to as the discretization parameter. As approximations of the system (3.3) we consider discrete problems of the form:
Find $x_h \in X_h$ such that:

$$a(x_h, y) = \langle f, y \rangle, \quad y \in X_h. \tag{3.7}$$

It is clear that the boundedness of the bilinear form a on X, given by (3.4), in particular implies boundedness on X_h. On the other hand, the inf–sup condition (3.5) will not imply that the corresponding condition holds on X_h. In general, the subspace X_h may even be constructed such that the system (3.7) is singular. However, the corresponding discrete inf–sup condition,

$$\inf_{x \in X_h} \sup_{y \in X_h} \frac{a(x,y)}{\|x\|_X \|y\|_X} \geq c_3 > 0, \tag{3.8}$$

where the constant c_3 is independent of h, will indeed ensure stability.

If the stability condition (3.8) holds then the system (3.7) can be written as a linear system of the form

$$\mathcal{A}_h x_h = f_h,$$

where $\mathcal{A}_h : X_h \to X_h^*$ is defined by

$$\langle \mathcal{A}_h x, y \rangle = a(u,y), \quad x,y \in X_h.$$

Furthermore, in the discrete case we define the canonical preconditioner, $\mathcal{B}_h : X_h^* \to X_h$, by

$$\langle \mathcal{B}_h f, y \rangle_X = \langle f, y \rangle, \quad y \in X_h. \tag{3.9}$$

By arguing exactly as above we obtain that

$$\kappa(\mathcal{B}_h \mathcal{A}_h) \leq c_1/c_3,$$

and as a consequence CGN, applied to the preconditioned system

$$\mathcal{B}_h \mathcal{A}_h x_h = \mathcal{B}_h f_h,$$

will converge with a convergence rate which can be bounded in terms of c_1 and c_3. We refer to [1, Section 5] for more details.

Remark 4.1. The canonical preconditioners defined from (3.9) will for most problems not lead to efficient preconditioners, since they typically will be composed of inverses of discrete differential operators that are expensive to evaluate. To obtain efficient preconditioners these operators have to be replaced by alternative *spectrally equivalent* operators, which can be evaluated cheaply. Here we refer to two symmetric and positive definite operators $\mathcal{B}_{1,h}, \mathcal{B}_{2,h} : X_h^* \to X_h$ as spectrally equivalent if there are constants α_0 and α_1, independent of the discretization parameter h, such that

$$\alpha_0 \langle f, B_{1,h} f \rangle \leq \langle f, B_{2,h} f \rangle \leq \alpha_1 \langle f, B_{1,h} f \rangle, \quad f \in X_h^*.$$

This is why we replaced the exact inverses by corresponding AMG operators in the numerical experiments above. We refer to [1, Section 6] for more details.

4 The Numerical Experiments – Revisited

From the discussion above we can conclude that if discrete linear systems of the form (3.7) are preconditioned by a canonical preconditioner of the form (3.9), then the condition number of the preconditioned system can be bounded in terms of the boundedness constant c_1 and the inf–sup constant c_3. Below we will revisit the examples we presented in Section 2 above, and discuss how the results we observed there can be explained by the theory we just have outlined.

Example 4.5. *The Stokes problem.*
The coefficient operator of the Stokes problem is of the form

$$\mathcal{A} = \begin{pmatrix} -\Delta & -\text{grad} \\ \text{div} & 0 \end{pmatrix},$$

and, as we stated above, this operator can be seen as an isomorphism mapping $X = (H_0^1(\Omega))^n \times L_0^2(\Omega)$ into $X^* = (H^{-1}(\Omega))^n \times L_0^2(\Omega)$, see e.g., [35]. The canonical preconditioner $\mathcal{B}$ should therefore be a Riesz mapping from X^* to X, and in this case this operator can be taken to be of the form

$$\mathcal{B} = \begin{pmatrix} -\Delta^{-1} & 0 \\ 0 & I \end{pmatrix}. \tag{4.1}$$

Furthermore, the corresponding discrete operator will be composed of the inverse of the discrete vector Laplacian, and a mass matrix (replacing the identity). It is well known that multigrid methods leads to spectrally equivalent and computational efficient analogs of these operators. Therefore, both the condition number and the number of iterations required by CGN, shown in Figure **(1)**, are bounded independently of the mesh parameter.

Example 4.6. *The linear elasiticy problem.*
In Example 4.2 we considered the mixed formulation of the linear elasticity problem. The coefficient operator of this problem is of the form,

$$\mathcal{A}_\varepsilon = \begin{pmatrix} -\Delta & -\text{grad} \\ \text{div} & -\varepsilon^2 I \end{pmatrix},$$

where $\varepsilon^2 = 1/(1+\lambda)$. It is easy to see that for $\varepsilon \in [0,1]$ the operator $\mathcal{A}_\varepsilon$ is an isomorphism mapping $X = (H_0^1(\Omega))^n \times L^2(\Omega)$ into $X^* = (H^{-1}(\Omega))^n \times L^2(\Omega)$, with operator norms bounded independently of ε. Therefore, the preconditioner used for the Stokes problem is also the appropriate canonical preconditioner here, and the condition number of the preconditioned problem is bounded uniformly in ε. This explains the uniform results, both with respect to the elasticity constant λ and the mesh parameter h, observed for the discrete version of the preconditioner $\mathcal{B}_1$ in Figure **(3)**.

Example 4.7. *The stabilized Stokes problem.*
The coefficient operator of the stabilized Stokes problem studied in Example 4.3 is given by

$$\mathcal{A}_\varepsilon = \begin{pmatrix} -\Delta & -\text{grad} \\ \text{div} & \varepsilon^2 \Delta \end{pmatrix}.$$

Above we studied finite element approximations of this operator when the parameter ε was proportional to the mesh paramer h. The motivation for such studies is the desire to stabilize discretizations of the Stokes problem, where the velocity and the pressure are approximated by finite element spaces of the same polynomial order. However, the system (2.6) also appears in different settings, for example through time-stepping schemes for simplifications of the Biot equations describing poroelastic problems c.f. e.g. [36, 37]. In these applications there is no relation between the positive parameter ε and the mesh parameter h. Therefore, we will discuss preconditioners for $\mathcal{A}_\varepsilon$ in the more general situation, where no relation between ε and h is assumed. In particular, we will derive preconditioners for $\mathcal{A}_\varepsilon$ in the continous case.
It is relatively straightforward to check that for each positive ε the operator $\mathcal{A}_\varepsilon$ is an isomorphism mapping $X = (H_0^1(\Omega))^n \times H^1(\Omega)$ onto $X^* = (H^{-1}(\Omega))^n \times H_0^{-1}(\Omega)$. Here, $H_0^{-1}(\Omega) \subset L_0^2(\Omega)$ represents

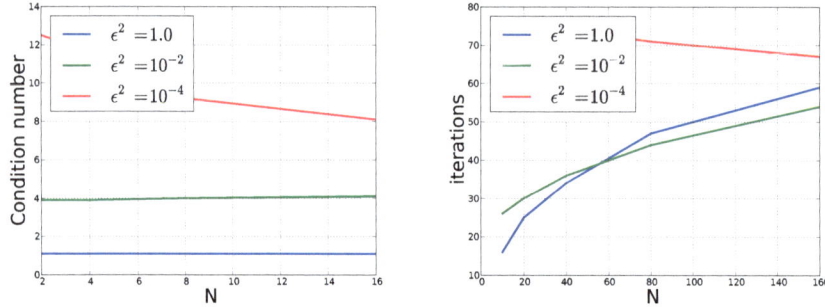

Figure (**7**): The left figure shows the condition number of the preconditioned system for the stabilized Stokes problem using Taylor–Hood elements, as a function of the mesh refinements and ε. The right figure shows the number of iterations required for CGN combined with an AMG preconditioner. The convergence criteria was a relative reduction of the preconditioned residual of a factor greater than 10^{10} in the discrete L_2 norm.

the dual of $H^1 \cap L_0^2$. However, the bounds on the associated operator norms on $\mathcal{A}_\varepsilon$ and $\mathcal{A}_\varepsilon^{-1}$ will depend on ε. In order to obtain corresponding ε–independent bounds on the operator norms, we need to introduce an ε–dependent norm on the solution space X, which degenerates to the L^2–norm as ε tends to zero. Let $X_\varepsilon = (H_0^1(\Omega))^n \times (L_0^2 \cap \varepsilon H^1)(\Omega)$, and X_ε^* the associated dual space. Here, we have used the notation of intersection of Hilbert spaces, which we will discuss in more details below. The space $(L_0^2 \cap \varepsilon \cdot H^1)(\Omega)$ is equal to $H^1(\Omega)$ as a set for each $\varepsilon > 0$, but its norm is given by

$$\|q\|_{L^2 \cap \varepsilon H^1}^2 = \|q\|_{L^2}^2 + \varepsilon^2 \|q\|_{H^1}^2.$$

For $\varepsilon = 0$ the space $(L_0^2 \cap \varepsilon \cdot H^1)(\Omega)$ coincides with $L_0^2(\Omega)$, with identical norms, and hence the mapping property of $\mathcal{A}_\varepsilon$ degenerates to the mapping property of the coefficient operator for Stokes problem as ε tends to zero.

The operator $\mathcal{A}_\varepsilon$ is an isomorphism mapping X_ε to X_ε^*, and with associated operator norms bounded independently of ε. As a consequence, if we use the Riesz mapping $\mathcal{B}_\varepsilon$ from X_ε^* to X_ε as a preconditioner, then the condition number $\kappa(\mathcal{B}_\varepsilon \mathcal{A}_\varepsilon)$ will be bounded independently of ε, cf. (3.6). The Riesz mapping from X_ε^* to X_ε is in the present case given by

$$\mathcal{B}_\varepsilon = \begin{pmatrix} -\Delta^{-1} & 0 \\ 0 & (I - \varepsilon^2 \Delta)^{-1} \end{pmatrix}.$$

In Figure (**7**) we provide numerical experiments for different values of ε and mesh size h. We use the Taylor–Hood element since we allow $\varepsilon \to 0$ independently of h. Clearly, the condition numbers remain bounded independently of both h and ε. The number of iterations required for convergence seems to be increasing for large ε, but this is probably because the asymptotic limit is not obtained for large ε and h. In fact, the number of iterations required in this problem is lower than that for the Stokes problem, see Example 4.1.

It can also be seen that if ε^2 is bounded by βh^2, where the constant β is independent of h, then the operators $\mathcal{B}_{0,h}$ and $\mathcal{B}_{\varepsilon,h}$ are spectrally equivalent. This explains the numerical results of Example 4.3, where we observes that the preconditioner $\mathcal{B}_{0,h}$ behaved uniformly well with respect to h for $\varepsilon^2 = \beta h^2$.

The two last examples we have encountered above both have a coefficient operator which depends on a parameter ε. For the linear elasticity problem we saw that we obtained a uniform conditioning of the preconditioned system by using an ε–independent preconditioner, while for the perturbed Stokes system (2.6) we use ε–dependent preconditioner to obtain uniform conditioning in ε, although a standard Stokes preconditioner works fine for $\varepsilon \approx h$ as seen in Example 4.3. The different behaviour of the two systems is reflected in the properties that the mixed linear elasticity problem is a regular perturbation problem, in the sense that the function spaces X and X^* remain unchanged as ε tends to zero, while the perturbed Stokes system (2.6) is a singular perturbation problem. For such problems we typically need to introduce parameter dependent function spaces in order to obtain parameter independent bounds on the coefficient operator and its inverse, and hence ε–independent conditioning of the preconditioned systems, cf. (3.6).

One way to introduce parameter dependent function spaces, which frequently occur in practice, is to consider weighted sums and intersections of Hilbert spaces. If X and Y are Hilbert spaces, then the intersection $X \cap Y$ is again a Hilbert space with norm

$$\|x\|_{X \cap Y}^2 = \|x\|_X^2 + \|x\|_Y^2.$$

Furthermore, if $\varepsilon > 0$ is a parameter then the corresponding weighted space, $X \cap \varepsilon \cdot Y$, is equal to $X \cap Y$ as a set for $\varepsilon > 0$ and to X for $\varepsilon = 0$, while the corresponding norms are given by

$$\|x\|_{X \cap \varepsilon \cdot Y}^2 = \|x\|_X^2 + \varepsilon^2 \|x\|_Y^2.$$

Hence, formally the norms behave continuously as ε tends to zero. In Example 4.7 we encountered an example of a space of this form.

Another related parameter dependent space is the weighted sum of Hilbert spaces. In general the space $X + Y$ consists of all elements $z = x + y$, $x \in X$, $y \in Y$, with norm given by

$$\|z\|_{X+Y}^2 = \inf_{\substack{z=x+y \\ x \in X, y \in Y}} (\|x\|_X^2 + \|y\|_Y^2),$$

while the corresponding weighted space, $X + \varepsilon^{-1} \cdot Y$, has the norm

$$\|z\|_{X+\varepsilon^{-1} \cdot Y}^2 = \inf_{\substack{z=x+y \\ x \in X, y \in Y}} (\|x\|_X^2 + \varepsilon^{-2} \|y\|_Y^2).$$

If $X \cap Y$ is dense in both X and Y, and $\varepsilon > 0$, then

$$(X \cap \varepsilon \cdot Y)^* = X^* + \varepsilon^{-1} \cdot Y^*,$$

where the star indicates dual spaces. Furthermore, there exist a corresponding Riesz mapping $R_\varepsilon : X^* + \varepsilon^{-1} \cdot Y^* \to X \cap \varepsilon \cdot Y$, such that the operator norms

$$\|R_\varepsilon\|_{L(X^*+\varepsilon^{-1} \cdot Y^*, X \cap \varepsilon \cdot Y)} \quad \text{and} \quad \|R_\varepsilon^{-1}\|_{L(X \cap \varepsilon \cdot Y, X^*+\varepsilon^{-1} \cdot Y^*)}$$

are bounded independently of ε. In fact, the operator R_ε is given by $z \mapsto x = R_\varepsilon z$, where $x \in X \cap Y$ solves the problem:

$$\langle x, y \rangle_X + \varepsilon^2 \langle x, y \rangle_Y = \langle f, y \rangle + \langle g, y \rangle, \quad y \in X \cap Y.$$

Here we have assumed that $z = f + g$ and that $f \in X^*$ and $g \in Y^*$ is chosen such that $\|z\|_{X^*+\varepsilon^{-1} \cdot Y^*}^2 = \|f\|_{X^*}^2 + \varepsilon^{-2} \|g\|_{Y^*}^2$. As above, we have used $\langle \cdot, \cdot \rangle$ to denote the proper duality pairings. We refer to [38] for more details on sums and intersections of Hilbert spaces.

The significance of the weighted sums and intersections of Hilbert spaces for singular perturbation problem is illustrated by the following simple example.

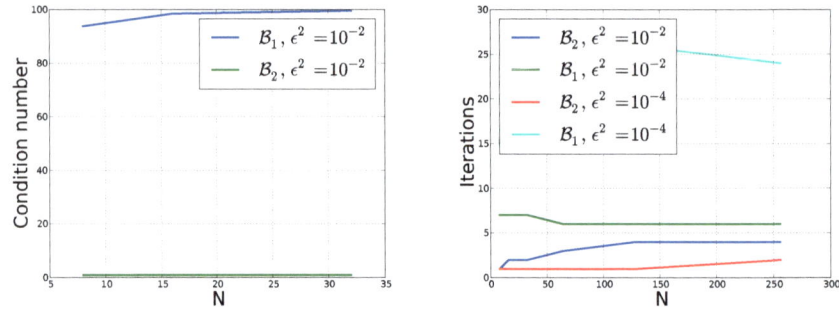

Figure **(8)**: The rightmost figure shows the condition number of the preconditioned matrix for different mesh resolutions. The leftmost figure shows the number of iterations required for convergence using CG combined with AMG, where the convergence criterion was a relative reduction of a factor 10^6 of the residual in the discrete L_2 norm.

Example 4.8. *Reaction–diffusion equation.*
Consider the reaction diffusion equation:

$$\mathcal{A}_\varepsilon u = u - \varepsilon^2 \Delta u \;=\; f, \text{ in } \Omega,$$
$$u \;=\; 0, \text{ on } \partial\Omega.$$

For each fixed positive ε the operator $\mathcal{A}_\varepsilon$ is a mapping from $H_0^1(\Omega)$ to $H^{-1}(\Omega)$, but it degenerates to the identity operator as ε tends to zero. Therefore, one possible choice for a preconditioner is,

$$\mathcal{B}_1 = -\Delta^{-1}.$$

However, this operator will not be an efficient preconditioner for ε close to zero. To obtain a precondi-tioner which is uniform with respect to ε we need to consider $\mathcal{A}_\varepsilon$ as an operator from $(L^2 \cap \varepsilon \cdot H_0^1)(\Omega)$ to its proper dual space $(L^2 + \varepsilon H^{-1})(\Omega)$, with associated Riesz mapping given by

$$\mathcal{B}_2 = (I - \varepsilon^2 \Delta)^{-1}.$$

Of course, the preconditioned operator satisfies $\mathcal{B}_2 \mathcal{A}_\varepsilon = I$, and using $\mathcal{B}_2$ involves solving the original reaction-diffusion problem. In practice, however, we replace $\mathcal{B}_2$ with equivalent and efficient opera-tors, where it is crucial that the equivalence is independent of both ε and the characteristic mesh size. In Figure **(8)** we compute the condition numbers of the corresponding preconditioned discrete systems using continuous piecewise linears finite elements to approximate the solution u. The precondition-ers are taken as discrete analogs of $\mathcal{B}_1$ and $\mathcal{B}_2$. We also show the number of iterations required by CG combined with AMG for a given convergence criteria. Clearly, as expected, we observe that the preconditioner $\mathcal{B}_2$ is superior to $\mathcal{B}_1$, in particular when ε is small.

Example 4.9. *Linear elasiticy in the primal variable.*
In Example 4.2 we saw that a standard multigrid method did not perform as a uniform preconditioner with respect to λ for the linear elasticity problem in primal variables. For $\mu = 1$ and $\varepsilon^2 = 1/(1+\lambda)$, this equation can be written as

$$\mathcal{A}_\varepsilon u = -\operatorname{grad}\operatorname{div} u - \varepsilon^2 \Delta u \;=\; \varepsilon^2 f, \quad \text{in } \Omega,$$
$$u \;=\; 0, \quad \text{on } \partial\Omega.$$

Here the coefficient operator $\mathcal{A}_\varepsilon$ can be seen as an isomorphism defined on $H_0(\mathrm{div},\Omega) \cap \varepsilon \cdot H_0^1(\Omega)^n$ to its proper dual space. So it easy to identify the proper canonical preconditioner at the continuous level as the operator

$$\mathcal{B}_\varepsilon = (-\,\mathrm{grad}\,\mathrm{div}\,-\varepsilon^2\Delta)^{-1}.$$

As in Example 4.8 we must replace $\mathcal{B}_\varepsilon$ with equivalent but efficient operators. However, standard multigrid algorithms do not produce preconditioners which behave uniformly with respect to ε for the most common finite element discretizations of this operator, and this is the reason for the poor performance in Example 4.2. Suitable multigrid methods are however described in [39].

Example 4.10. *Time dependent Stokes problem.*
In [32] we showed that

$$\mathcal{A}_\varepsilon = \begin{pmatrix} I - \varepsilon^2\Delta & -\,\mathrm{grad} \\ \mathrm{div} & 0 \end{pmatrix}$$

was a bounded and continuously invertible linear operator mapping $X_\varepsilon = (L^2 \cap \varepsilon H^1)(\Omega) \times ((H^1 \cap L_0^2) + \varepsilon L_0^2)(\Omega)$ onto its dual space. The preconditioner

$$\mathcal{B}_\varepsilon = \begin{pmatrix} (I - \varepsilon^2\Delta)^{-1} & 0 \\ 0 & (-\Delta)^{-1} + \varepsilon^2 I \end{pmatrix}$$

corresponds to a canonical mapping from the dual space and back to X_ε. This explains the efficiency of the preconditioner studied in Example 4.4.
We remark that an alternative preconditioner at the continuous level is

$$\mathcal{B}_\varepsilon = \begin{pmatrix} (I - \mathrm{grad}\,\mathrm{div}\,-\varepsilon^2\Delta)^{-1} & 0 \\ 0 & I \end{pmatrix}.$$

This follows since $\mathcal{A}_\varepsilon$ can be seen as an isomorphism mapping $H(\mathrm{div},\Omega) \cap \varepsilon H^1(\Omega) \times L^2(\Omega)$ onto its dual space, and with appropriate operator norms bounded independently of ε. For discussions of discrete analogs of this preconditioner we refer to [12, 13, 39].

5   Conclusions

In this chapter we have described how preconditioners for systems of partial differential equations can be constructed based on the mapping properties of the differential operator in properly chosen Sobolev spaces. In particular, the canonical preconditioners can be seen as Riesz isomorphisms mapping the space of the right hand side into the solution space. These Riesz isomorphisms are then replaced by equivalent and efficient precondititioners, constructed by using, for example, multigrid or domain decomposition techniques.

Bibliography

[1] Mardal KA, Winther R, Preconditioning discretizations of systems of partial differential equations, Numer Lin Alg Appl 2011; 18(1):1–40.

[2] Arnold DN, Falk RS, and Winther R, Preconditioning discrete approximations of the Reissner-Mindlin plate model, RAIRO Modél Math Anal Numér 1997; 31(4): 517–557.

[3] Arnold DN, Falk RS, and Winther R, Preconditioning in H(div) and applications, Math Comput 1997;66:957–984.

[4] Axelsson O and Karatson J, Equivalent operator preconditioning for elliptic problems, Numer Algor 2009; 50:297–380.

[5] Faber V, Manteuffel T and Parter SV, On the theory of equivalent operators and application to the numerical solution of uniformly elliptic partial differential equations, Adv Appl Math 1990;11: 109–163.

[6] Hiptmair R, Operator preconditioning, Comput Math Appl 2006;52(5):699–706.

[7] Kirby RC, From functional analysis to iterative methods, Accepted in SIAM Review, 2010.

[8] Mardal KA, Nilssen TK, and Staff GA, Order optimal preconditioners for implicit Runge-Kutta schemes applied to parabolic PDEs, SIAM J Sci Comput 2007;29(1): 361–375.

[9] Staff GA, Mardal KA, and Nilssen TK, Preconditioning of fully implicit Runge-Kutta schemes for parabolic PDEs, Modeling, Identification and Control 2006;27(2): 109–123.

[10] Benzi M, Golub GH, and Liesen J, Numerical solution of saddle point problems, Acta Numer 2005;14:1–137.

[11] Benzi M and Liu J, Block preconditioning for saddle point systems with indefinite $(1,1)$ block, Int J Comput Math 2007;84(8): 1117–1129.

[12] Benzi M and Olshanskii MA, An augmented Lagrangian-based approach to the Oseen problem, SIAM J Sci Comput 2006; 28:2095–2113.

[13] Benzi M, Olshanskii MA, and Wang Z, Augmented Lagrangian preconditioners for the incompressible Navier-Stokes equations, url:http://www.mathcs.emory.edu/ molshan/OLSH/publ.html.

[14] Elman HC, Silvester DJ and Wathen, AJ, Finite elements and fast iterative solvers: with applications in incompressible fluid dynamics, Numerical Mathematics and Scientific Computation, Oxford University Press, New York; 2005.

[15] Klawonn A, Block-triangular preconditioners for saddle point problems with a penalty term, SIAM J Sci Comput 1998;19(1):172–184.

[16] Klawonn A and Starke G, Block triangular preconditioners for nonsymmetric saddle point problems: field-of-values analysis, Numer Math 1999;81:577–594.

[17] Loghin D and Wathen AJ, Analysis of preconditioners for saddle-point problems, SIAM J Sci Comput 2004;25: 2029–2049.

[18] Olshanskii MA and Vassilevski Y, Pressure Schur complement preconditioners for the discrete Oseen problem, SIAM J Sci Comput 2007; 29:2686–2704.

[19] FEniCS – finite element software package, URL: http://www.fenics.org, 2010.

[20] Mardal KA, Block preconditioning of systems of PDEs. In Logg A, Mardal K.-A. , and Wells G, Eds, Automated Scientific Computing. Springer-Verlag; In Press.

[21] Heroux MA, Bartlett RA, Howle VE *et al*, An overview of the trilinos project, ACM Trans. Math Softw. 2005;31(3):397–423.

[22] Rusten T and Winther R, A preconditioned iterative method for saddlepoint problems, SIAM J Matrix Anal Appl 1992;13(3):887–904,

[23] Braess D, Finite elements; theory, fast solvers, and applications in solid mechanics, Cambridge University Press, 1997.

[24] Brenner SC and Scott LR, The mathematical theory of finite element methods, Text in Applied Mathematics, Springer-Verlag, 1994.

[25] Engelman MS, Sani RI, Gresho PM *et al*, Consistent vs. reduced integration penalty methods for incompressible media using several old and new elements, Int J Num Meth in Fluids 1982;2:25–42.

[26] Hughes TJR, Liu WK, and Brooks A, Finite element analysis of incompressible viscous flows by the penalty function formulation, J Comp Phys 1979;30:1–60.

[27] Mardal KA, Tai XC, and Winther R, A robust finite element method for Darcy–Stokes flow, SIAM J Numer Anal 2002;40:1605–1631.

[28] Hughes TJR and Franca LP, A new finite element formulation for computational fluid dynamics: VII. The Stokes problem with various well-posed boundary conditions: symmetric formulations that converge for all velocity/pressure spaces, Comput Meth Appl Mech Engnrg 1987;65:85–96.

[29] Hughes TJR, Franca LP, and Balestra M, A new finite element formulation for computational fluid dynamics: V. Circumventing the Babuska-Brezzi condition: a stable Petrov-Galerkin formulation of the Stokes problem accommodating equal-order interpolations, Comput Meth Appl Mech Engnrg 1986;59:85–99.

[30] Cahouet J and Chabard JP, Some fast 3D finite element solvers for the generalized Stokes problem, Int J Numer Meth Fluids 1988;8(8):869–895.

[31] Turek S, Efficient solvers for incompressible flow problems, Springer; 1999.

[32] Mardal KA, Winther R, Uniform preconditioners for the time dependent Stokes problem, Numer Math 2004; 98(2): 305–327. Erratum in: Numer Math 2006; 103(1): 171–172.

[33] Babuška I, Error bounds for finite element methods, Numer Math 1971;16:322–333.

[34] Babuška I and Aziz AK, Survey lectures on the mathematical foundation of the finite element method, The Mathematical Foundation of the Finite Element Method with Applications to Partial Differential Equations (AK. Aziz, ed.), Academic Press, New York - London 1972; 3–345.

[35] Brezzi F and Fortin M, Mixed and hybrid finite element methods, Springer-Verlag; 1991.

[36] Aguilar G, Gaspar F, Lisbona F, and Rodrigo C, Numerical stabilization of Biot's consolidation model by a perturbation on the flow equation, Int J Numer Meth Engng 2008;75:1282–1300.

[37] Biot MA, General theory of three-dimensional consolidation, J Appl Physics 1941;12:155–164.

[38] Bergh J and Löfström J, Interpolation spaces, Springer-Verlag, 1976.

[39] Schöberl J, Multigrid methods for a parameter dependent problem in primal variables, Numer Math 1999;84:97–119.

Chapter 5

Automatic Construction of Sparse Preconditioners for High-Order Finite Element Methods

Travis M. Austin[1], Marian Brezina[2], Thomas A. Manteuffel[3] and John Ruge[4]

Abstract: High-order finite elements are necessary for many large scale scientific computing problems because either long time integrations need high accuracy at each time step or modelling certain physics requires the use of high-order finite elements. To reduce the costs of high-order finite elements, robust and efficient linear solvers with minimal memory requirements are crucial. This implies the need for a robust and efficient preconditioner. In the 1980s, Orszag [J. Comp. Phys, 37, 70 (1980)] showed that an inexpensive preconditioner can be defined based on a matrix that is generated from rediscretizing the finite element problem on a finer mesh using low-order finite elements. The matrix was sparse and its inverse proved to be an effective preconditioner.

In this work, we describe a similar sparse preconditioner. However, this preconditioner does not require the user to rediscretize the problem, but only to provide access to the high-order element stiffness matrices. The preconditioner is then an approximate inversion of this sparse matrix with algebraic multigrid, or any other robust method. The sparse matrix is constructed by representing the high-order element stiffness matrices with a sparse approximation generated using a least-squares approach. The least-squares approach aims to accurately represent the low-order modes of the high-order element stiffness matrix. We compare results to Orszag's approach, where the inverse of the sparse matrix is approximated with a single V-cycle of algebraic multigrid, and to a naive approach that uses a single V-cycle of algebraic multigrid, built from the original high-order system, as the preconditioner.

Keywords: *partial differential equation, discretization, high-order finite elements, preconditioning, sparse matrix, robust, algebraic multigrid*

[1] Tech-X Corporation, Boulder, CO 80303; e-mail: austin@txcorp.com

[2] Department of Applied Mathematics, University of Colorado, Boulder, USA; e-mail: mbrezina@math.cudenver.edu

[3] (corresponding author) Department of Applied Mathematics, University of Colorado, Boulder, USA; e-mail: tmanteuf@colorado.edu

[4] Department of Applied Mathematics, University of Colorado, Boulder, USA; e-mail: jruge@colorado.edu

Owe Axelsson and János Karátson (Eds)

1 Introduction

The robustness of an iterative method for solving large sparse linear systems generated from finite element discretizations depends on the robustness of the preconditioner. For Krylov methods a good preconditioner reduces the condition number of the matrix defining the Krylov subspace so only a few iterations are needed to obtain an accurate solution. To construct an efficient iterative solver, the preconditioner needs to be designed to ensure that its application is not too costly.

For the hierarchical p-version finite element method, a number of preconditioners have been examined that split the system into a low-order part and a high-order part [1–3]. Preconditioners based on either the block diagonal or approximate Schur complements have been proposed that have been shown to be robust in p and h. Here, we focus on the spectral element method [4] and aim for a preconditioner that is more black box so that existing spectral element codes can easily use the new preconditioner through an external library.

To that end, we consider previous preconditioning efforts for the spectral element method. In the past, particular focus has been on Schwarz-like methods that require large local subdomain solves [5–8]. As was reported by Widlund in [6,9] and Mandel in [5], a coarse grid solver that couples the subdomains is necessary to maintain optimal convergence. However, this optimality requires solving the dense local blocks with direct methods and performing a coarse-grid solve that can be relatively expensive. For certain problems, some of the high computational cost associated with such methods has been overcome [10–12]. For a comparison between Schwarz-based methods and multigrid, we refer the reader to [13].

In the 1980s, Orszag discovered that a low cost preconditioner can be built from a sparser, lower-order version of the high-order finite element matrix [14]. This approach has also been investigated by a number of other researchers in [15–19]. The sparse matrix is constructed by discretizing the finite element problem using low-order finite elements on a mesh defined by the high-order finite element nodes. The preconditioner is then defined to be an approximate inversion of this sparse matrix. Automating the construction of the sparse matrix so that an actual rediscretization is not required is attractive since it can reduce the effort to employ such a sparse preconditioner in a finite element code.

A novel approach to generating a lower-order sparse matrix for use in the preconditioning is presented in this work. The approach is partially motivated by [20]. A key component of the approach is the approximate inversion of the sparser matrix in the preconditioning stage with algebraic multigrid (AMG). Furthermore, the approach does not require a full rediscretization involving numerical integration with low-order finite elements, but instead requires solving a local least-squares problem defined by the high-order finite element stiffness matrix. With recent advances in multicore architectures, the time to setup and solve the local least-squares problem can be minimal. The resulting sparse element stiffness matrices are used in a finite element assembly process to generate an overall sparser representation to the global stiffness matrix. We compare performance of the two low-order preconditioners to an AMG preconditioner constructed using the high-order system.

For elliptic problems, we will see that the automatic sparse approach consistently produces better performance than the low-order discretization. These differences are more distinct when using Chebyshev-Legendre nodes than when using Gauss-Legendre-Lobatto nodes for the numerical integration of the matrix coefficients. As the polynomial order goes from 2nd- to 10th-order in 2D, we see that both sparse approaches also outperform an AMG preconditioner built from the high-order system. These differences in performance arise when we account for the cheaper application of the sparser preconditioners. For example, for 10th-order polynomials in 2D, an application of the AMG preconditioner built from the sparse matrix is nearly a tenth of the cost of an application of the AMG preconditioner built from the high-order system. Less extensive 3D results reconfirm the conclusions reached from the 2D results.

The remainder of the paper proceeds as follows. In Section 2 we present the discretization approach with a brief discussion on numerical quadrature approaches. We then describe in Section 3 the process of discretizing using low-order finite elements. In Section 4 the least-squares approach for constructing a sparse preconditioner is described. Computational results for the least-squares approach along with the low-order discretization approach are mentioned in Section 5, in addition to approximate condition numbers for the preconditioned systems. The complete discussion of computational results takes place in Section 6. We end with concluding remarks in Section 7.

2 Discretization Approach

For $x \in \mathbf{R}^d$ with $d = 2, 3$, the model problem used for describing the discretization approach is Poisson's equation:[5]

$$\begin{aligned} -\nabla \cdot \nabla \phi &= f \quad \forall x \in \Omega \\ \phi &= 0 \quad \forall x \in \partial\Omega. \end{aligned} \tag{2.1}$$

For simplicity, we consider the homogeneous Dirichlet case here and note that this work is easily extended to the inhomogeneous case. Next, let $V_0^h := span(v_1, v_2, ..., v_n) \subset H_0^1(\Omega)$ such that $v_i|_{\partial\Omega} = 0$ for $i = 1, ..., n$, and define the weak form: find $\phi_h \in V_0^h$ s.t.

$$a(\phi_h, v_h) = (f, v_h) \quad \forall v_h \in V_0^h, \tag{2.2}$$

where

$$a(\phi_h, v_h) := \int_\Omega \nabla \phi_h \cdot \nabla v_h \, dx, \qquad (f, v_h) := \int_\Omega f v_h \, dx.$$

The weak form yields the matrix problem

$$\mathbf{A}_h \mathbf{x} = \mathbf{b},$$

where

$$\mathbf{A}_h = [a(v_i, v_j)]_{i,j=1}^n, \qquad \mathbf{b} = [(f, v_i)]_i^n.$$

We define the high-order finite element method as one that uses a basis, V_h, that consists of polynomials of degree $p > 1$. In this work, we consider nodal basis functions defined as Lagrangian interpolants such that $v_i(\xi_j) = \delta_{ij}$. The shapes of the basis functions are defined by the nodal arrangement on the quadrilateral or hexahedral element. While equally spaced nodes yield the desired accuracy, the resulting system of equations can be extremely ill-conditioned, particularly if $p > 4$ [21–23].

A common choice for nodal arrangements (cf. [22, 23]) relies on the Gauss-Legendre-Lobatto (GLL) points, which are the roots of

$$(1 - \xi_j^2) L_n'(\xi_j) = 0, \quad j = 0, ..., N,$$

where $L_n(x)$ is the nth Legendre polynomial for the 1D element, $[-1, 1]$. For 2D and 3D elements, appropriate tensor-product forms of these locations are employed as in Fig. (**1**). Also used for nodal

---

[5]In this work, the serial and parallel FOSPACK codes are used to discretize 2D and 3D Poisson's equation using the appropriate interior and boundary functional. Here, we set $f = 0$ and apply homogeneous Dirichlet boundary conditions through a boundary functional. The inhomogeneous case is treated via a first-order systems least-squares formulation of Poisson's equation.

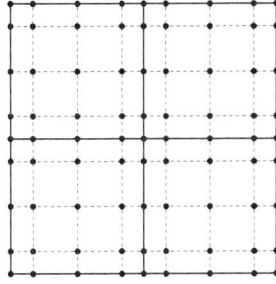

Figure **(1)**: Gauss nodes and mesh (solid black) for four biquartic element on square element overlaid with a high resolution mesh (dashed black) discretized with bilinear elements on Gauss nodes.

arrangements are the Chebyshev-Legendre (CL) points, given by $\xi_j = \cos(\pi j/N)$, for $0 \leq j \leq N$. GLL and CL points have similar locations and both have significantly better conditioning than for equally spaced nodes [21]. CL points have the advantage that they are nested since $\cos(2\pi k/2N) = \cos(\pi k/N)$.

When using GLL points for the node locations, the same points are often used for quadrature. This produces a diagonal approximation to the mass matrix and a very efficient matrix-vector product for the stiffness matrix. However, it also introduces quadrature error into the mass matrix, and in more than one dimension, the stiffness matrix. In [24] the effect of this error on convergence was examined. For a regular mesh, no reduction in convergence due to reduced integration was observed, but for a nonuniform mesh, a reduction in convergence due to the approximation of the mass matrix was seen. Finally, it was noted that using GLL quadrature to compute the stiffness matrix reduces computation time by a half over using exact numerical integration.

The 2D spectral element stiffness matrix for a quadrilateral element can be expressed as the Kronecker product

$$\mathbf{A}_e := \mathbf{M}_y \otimes \mathbf{A}_x + \mathbf{A}_y \otimes \mathbf{M}_x, \tag{2.3}$$

where the subscript x or y indicates the 1D elements in the given direction. Using GLL points for both the quadrature and node locations yields

$$\mathbf{A}_e := \tilde{\mathbf{M}}_y \otimes \mathbf{A}_x + \mathbf{A}_y \otimes \tilde{\mathbf{M}}_x, \tag{2.4}$$

where the approximate mass matrix, $\tilde{\mathbf{M}}$, is diagonal because of the GLL points being coincident with the node locations. In [13] Lottes and Fischer noted the particular advantage to expressing stiffness matrices in the form of Eqs. (2.3) and (2.4). Fast matrix-free algorithms based on Kronecker products can be used to speed-up computations and reduce memory. Success in using this approach can also seen in [19]. Further discussion on this topic is, however, beyond the scope of this paper.

Finally, we note that a bound on the L^2 norm of the error can be expressed using standard isoparametric approximation theory [25] as

$$\|\phi - \phi_h\|_{0,\Omega} \leq Ch^p \|\phi\|_{p,\Omega},$$

where ϕ is the exact solution of (2.1) and ϕ_h is the weak solution of (2.2). Thus, if $\phi \in H^p(\Omega)$ and if the polynomial order is at least $p - 1$, we can achieve $O(h^p)$ convergence. Such potential high-accuracy has led researchers to use high-order finite elements for a number of problems.

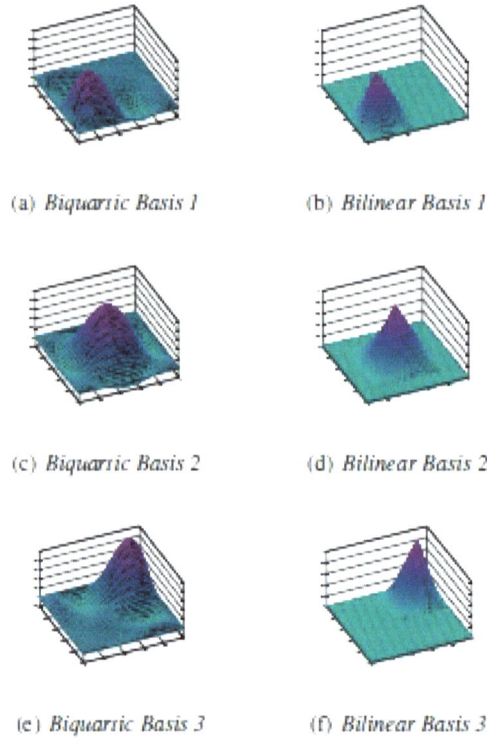

(a) *Biquartic Basis 1*

(b) *Bilinear Basis 1*

(c) *Biquartic Basis 2*

(d) *Bilinear Basis 2*

(e) *Biquartic Basis 3*

(f) *Bilinear Basis 3*

Figure **(2)**: Three of the 25 biquartic basis functions (left) and three of the 25 bilinear basis functions on the mesh created from the biquartic nodes (right). Biquartic basis functions (a) and (e) are coupled but the respective bilinear basis functions, (b) and (f), are uncoupled.

3   Low-order Sparse Matrix Construction

The concept of preconditioning a finite element problem employing high-order polynomials with a finite element problem employing low-order polynomials was first proposed in the 1980s by Orszag in [14]. He proposed using the high-order finite element nodes (cf. Fig. **(1)**) to define a *finer* finite element mesh and then employ bilinear finite elements on the finer mesh ensuring the resulting matrix has equal dimension to the matrix produced using the high-order polynomials. Parter in [26] and Parter and Kim in [17] have also explored such an approach in 1D and presented bounds on the low-order operator as a preconditioner for the high-order operator that is independent of h and p. Fischer has also examined this approach in [10] and argued for constructing triangular subelements instead of rectangular subelements. Heys *et al.* in [18] explored approximating the inversion of the low-order matrix in the preconditioning with AMG, while Olson in [27] explored this approach for the spectral element method on triangles. Brown in [19] also examined this approach in the context of a nonlinear solve. In Fig. **(2)** we have displayed 3 of the 25 biquartic basis functions next to the corresponding bilinear basis functions on the finer mesh. This figure illustrates the decoupling of *Bilinear Basis 1* and *3* in contrast to coupled *Biquartic Basis 1* and *3*, which overlap. The matrix produced by the bilinear basis functions will be sparser because of the reduced couplings.

The bilinear basis functions define a new space, $W_0^h := (w_1, w_2, ..., w_n) \subset H_0^1(\Omega)$ such that $w_i|_{\partial\Omega} = 0$

for $i = 1, ..., n$, and a new weak form: find $\psi_h \in W_0^h$ s.t.

$$a(\psi_h, w_h) = (f, w_h), \quad \forall w_h \in W_0^h.$$

We then get a new set of matrix equations, given by

$$\mathbf{A}_\ell \mathbf{y} = \mathbf{c},$$

where

$$\mathbf{A}_\ell = [a(w_i, w_j)]_{i,j=1}^n, \qquad \mathbf{c} = [(f, w_i)]_i^n.$$

Using this low-order operator in a preconditioning yields

$$\mathbf{A}_\ell^{-1} \mathbf{A}_h \mathbf{x} = \mathbf{A}_\ell^{-1} \mathbf{b}. \tag{3.1}$$

If GLL is used for both the node locations and the quadrature points for the high-order operator, $\mathbf{A}_h$, then, as noted by Lottes and Fischer in [13], the low-order operator, $\mathbf{A}_\ell$, should also be generated using GLL for quadrature points and node locations. In this work, when using GLL quadrature and node locations for high-order finite elements, we use the appropriate GLL for the low-order operator. For CL points, we perform exact quadrature for both the original system and the system used in the preconditioner. We compare the computational performance for various quadratures in Section 5.

4   Least-squares Sparse Matrix Construction

In Section 3, we described an approach to constructing a sparser preconditioner that required a rediscretization of the finite element problem. In this section, we describe an approach that eliminates the need for a rediscretization by basing the sparse preconditioner solely on the algebraic nature of the element stiffness matrices. No local geometric information is needed. For symmetric matrices, we describe the construction of a sparse matrix where it is assumed that the user can provide the existing local high-order finite element stiffness matrices. This was also the basis for the AMGe method presented in [20] and is normally information that is readily available with minor changes to a finite element code.

To this end, consider the weak form in Eq. (2.2) posed on the high-order finite element mesh in Fig. **(1)** and denote the collection of element stiffness matrices by $\mathcal{A}^h := \left\{ \mathbf{A}_{e_1}^h, \mathbf{A}_{e_2}^h, \mathbf{A}_{e_3}^h, \mathbf{A}_{e_4}^h \right\}$. Disregarding boundary conditions, each element stiffness matrix is completely dense with 625 nonzero entries. We want a collection of the same number of matrices given by $\mathcal{A}^\ell := \left\{ \mathbf{A}_{e_1}^\ell, \mathbf{A}_{e_2}^\ell, \mathbf{A}_{e_3}^\ell, \mathbf{A}_{e_4}^\ell \right\}$, where each $\mathbf{A}_{e_i}^\ell$ is allowed a sparsity equivalent to using bilinear finite elements on the mesh defined by the dashed black lines in Fig. **(1)**. Each $\mathbf{A}_{e_i}^\ell$ will have 169 nonzeros. With lexicographical ordering of the elements in Fig. **(1)**, the sparsity pattern of $\mathbf{A}^\ell$ will be as in Fig. **(3)** compared to being completely dense.

We denote the eigenspace of each $\mathbf{A}_{e_i}^h$ by $eig(\mathbf{A}_{e_i}^h) := \{(\mathbf{s}_k, \lambda_k)\}_{k=1}^{25}$, where $\mathbf{A}_{e_i}^h \mathbf{s}_k = \lambda_k \mathbf{s}_k$ for $k = 1, 25$. For simplicity, we denote $\mathbf{A}_{e_i}^h$ by $\mathbf{A}_i$ and define $\mathbf{C}_i$ to be a symmetric matrix of the same rank as $\mathbf{A}_i$ with the same nonzero structure as Fig. **(3)**. We then solve the minimization problem: Find $\mathbf{C}_i$ that minimizes

$$\sum_{\mathbf{s}_k \notin \mathcal{N}(\mathbf{A}_i)} \frac{1}{\lambda_k^2} \|\lambda_k \mathbf{s}_k - \mathbf{C}_i \mathbf{s}_k\|_0^2$$

subject to the constraint $\mathbf{C}_i = \mathbf{C}_i^T$ and $\mathbf{C}_i \mathbf{s}_k = 0$ for all $\mathbf{s}_k \in \mathcal{N}(\mathbf{A}_i)$. Here $\mathcal{N}(\mathbf{A}_i)$ denotes the nullspace of $\mathbf{A}_i$. The scaling is chosen to better approximate the lowest modes with the sparse approximation to the high-order stiffness matrix.

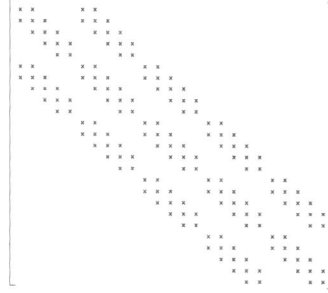

Figure (**3**): Sparsity pattern for bilinear elements on a single element of the mesh in Fig. (**1**) for which the corresponding biquartic element stiffness matrix is entirely dense.

To solve the least-squares problem, we construct a vector, z_i, denoting the set of coefficients in C_i that defines the sparse matrix approximation to A_i. We only solve for the lower-triangular nonzeros in C_i and use the symmetry to define the rest of C_i in order to enforce $C_i = C_i^T$. We then introduce a matrix, G_k, so that $G_k z_i$ generates a vector equivalent to $C_i s_k$ for each eigenvector s_k. This is merely re-arranging the representation of the unknown from a matrix, C_i, to a vector, z_i. We then have a set of matrix equations for z_i given by

$$\mathbf{G}z_i = \begin{pmatrix} \frac{1}{\lambda_1}\mathbf{G}_1 \\ \frac{1}{\lambda_2}\mathbf{G}_2 \\ \cdot \\ \cdot \\ \cdot \\ \frac{1}{\lambda_{n-1}}\mathbf{G}_{n-1} \end{pmatrix} \quad z_i = \begin{pmatrix} s_1 \\ s_2 \\ \cdot \\ \cdot \\ \cdot \\ s_{n-1} \end{pmatrix} = \mathbf{s},$$

subject to the constraint $\mathbf{H}z_i = 0$ where $\mathbf{H}$ is the null-space operator. Solving this set of constrained equations using the least-squares method is equivalent to solving the following system for the coefficients, z_i, and the Lagrange multipliers, γ_i:

$$\begin{pmatrix} \mathbf{G}^T\mathbf{G} & \mathbf{H}^T \\ \mathbf{H} & 0 \end{pmatrix} \begin{pmatrix} z_i \\ \gamma_i \end{pmatrix} = \begin{pmatrix} \mathbf{G}^T\mathbf{s} \\ 0 \end{pmatrix}. \tag{4.1}$$

Having z_i then defines the sparse element stiffness matrix, C_i, at which point we can build the global, sparse preconditioning matrix according to

$$\mathbf{A}_\ell = \sum_i \mathbf{P}_i \mathbf{C}_i \mathbf{P}_i^T,$$

given that

$$\mathbf{A}_h = \sum_i \mathbf{P}_i \mathbf{A}_i \mathbf{P}_i^T$$

and $\mathbf{P}_i$ is each element's local-to-global map. After assembly of $\mathbf{A}_\ell$, boundary conditions can be applied to ensure $\mathbf{A}_h$ and $\mathbf{A}_\ell$ treat boundaries equivalently. The preconditioned system is then given by

$$\mathbf{A}_\ell^{-1}\mathbf{A}_h x = \mathbf{A}_\ell^{-1}\mathbf{b}. \tag{4.2}$$

Here, $\mathbf{A}_\ell^{-1}$ is approximated by a single V-cycle of AMG constructed using the matrix $\mathbf{A}_\ell$. However, any other fast solver for $\mathbf{A}_\ell^{-1}$ will yield an effective algorithm.

A key component of this method is the determination of the sparsity pattern for $\mathbf{C}_i$ that is required before solving the least-squares problem. In the simplest case, we have information regarding the arrangement of nodes on the high-order element and know a node's nearest neighbors. We expect to have this information in most cases and can use it to define the nonzero pattern. However, we are currently exploring algebraic ways of selecting a sparsity pattern that does not require any knowledge of the nodal geometry within an element. Preliminary results on uniform meshes suggest this approach is viable as an alternative to having geometric information. For instance, long distance couplings in a high-order element stiffness matrix generally have small values relative to the diagonal and can be dropped from the sparsity pattern, as they would be when taking into account geometric information.

5   Computational Results

5.1   Problem types

For the computational results, we focus on elliptic problems as well as a First-Order Systems Least-Squares (FOSLS) formulation of the diffusion equation [28]. For the elliptic problems, we examine 2D and 3D Poisson's equation on a domain resulting from an approximately 27 degree angle of shear of a unit square or unit cube domain in the x-direction. We refer to these domains as the sheared square and sheared cube domains. We consider iteration counts, work estimates, and memory costs for finding the solution with the various approaches.

Furthermore, for 2D, we examine both GLL points used to define the nodes as in Fig. (**1**) and also CL points. For the CL points, exact integration is used for the high-order finite element matrix and for the low-order matrix. For GLL points, we use GLL points for node location as well as for quadrature, leading to reduced numerical integration.

For each elliptic problem, we impose homogeneous Dirichlet boundary conditions and a homogeneous source but with a random initial guess. We investigate the number of iterations required to reduce the norm of the residual by a factor of 10^{-10}. We also address memory usage and computational cost. In Eqs. (3.1) and (4.2), $\mathbf{A}_\ell^{-1}$ is approximated with a single AMG V(1,1)-cycle.

For 2D, we use serial AMG1r6, and for 3D, we use HYPRE's parallel BoomerAMG [29] on 1-8 processors. We use a strength of connection parameter of 0.25 for both AMG solvers leaving the remaining AMG parameters at their default values. We also compute approximate condition numbers found through calculating $\mathbf{A}_\ell^+\mathbf{A}_h$ on a single element for the sheared square and sheared cube domain, given that $\mathbf{A}_\ell^+$ is the pseudoinverse of $\mathbf{A}_\ell$. We calculate these approximate condition numbers numerically in Python using the NumPy package.

The FOSLS formulation of the diffusion equation permits investigation of the method for a realistic problem with an inhomogeneous source. The FOSLS formulation requires reposing Eq. (2.1) as a system of first-order equations, i.e.,

$$\begin{aligned}
\mathbf{u} + \nabla\phi &= \mathbf{0} \\
\nabla\cdot\mathbf{u} &= f,
\end{aligned} \tag{5.1}$$

with appropriate boundary conditions on $\mathbf{u}$ and ϕ. The standard FOSLS formulation then adds the additional constraint that says that $\mathbf{u}$ is irrotational. The system becomes

$$\begin{aligned}
\mathbf{u} + \nabla\phi &= \mathbf{0} \\
\nabla\cdot\mathbf{u} &= f \\
\nabla\times\mathbf{u} &= \mathbf{0}.
\end{aligned}$$

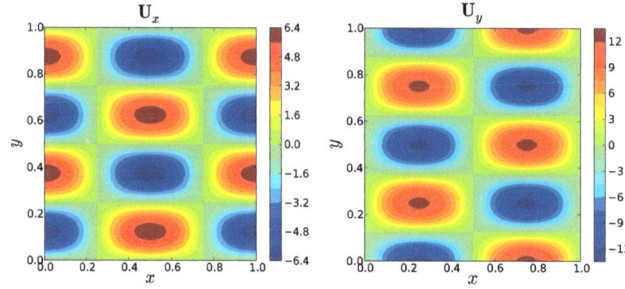

Figure **(4)**: Discrete solution using 4th-order finite elements to the FOSLS formulation of the diffusion equation posed with the functional defined by Eq. (5.2). The exact solution of the FOSLS formulation is defined by Eqs. (5.1) and (5.3).

We define the first least-squares functional:

$$\mathcal{F}_1(\phi, \mathbf{u}; f) := \|\mathbf{u} + \nabla\phi\|_0^2 + \|\nabla \cdot \mathbf{u} - f\|_0^2 + \|\nabla \times \mathbf{u}\|_0^2$$

and the second least-squares functional:

$$\mathcal{F}_2(\mathbf{u}; f) := \|\nabla \cdot \mathbf{u} - f\|_0^2 + \|\nabla \times \mathbf{u}\|_0^2 \tag{5.2}$$

in which we only solve for the flux, $\mathbf{u}$, getting the potential, ϕ, with a post-solve by minimizing the third functional

$$\mathcal{F}_3(\phi; \mathbf{u}) := \|\mathbf{u} - \nabla\phi\|_0^2.$$

For the FOSLS problem, we focus on $\mathcal{F}_2$ on a non-sheared domain and skip the post-solve for the potential, ϕ, involving $\mathcal{F}_3$. The domain is $\Omega := [0,1]^2$ and the source and boundary conditions are defined in terms of the prescribed potential, given by

$$\phi(x, y) = \sin(2\pi x)\sin(4\pi y), \tag{5.3}$$

and the flux, $\mathbf{u}(x, y)$, is defined by Eq. (5.1). In Fig. **(4)**, we illustrate the solution for U_x and U_y on a 64×64 mesh for 4th-order finite elements.

In the last column of Table **1**, we see the functional norm value defined by Eq. (5.2) for a fixed mesh and for increasing polynomial order. The decrease in the functional norm is seen as the polynomial order is increased. We also see in the second column (NIT_{HO}) the number of PCG iterations required to reduce the residual by a factor of 10^{-10} for the various polynomial orders using CG with an AMG preconditioner built directly from the high-order finite element system. In the third column (NIT_{LO}), we observe the number of PCG iterations required to reduce the residual by the same order using CG with an AMG preconditioner built from the sparse optimized low-order finite element system. Later, we investigate work differences that take into account the cost of a single iteration.

5.2 Condition number estimates

For the low-order system, $\mathbf{A}_\ell$, introduced in Section 2, the condition number of $\mathbf{A}_\ell^{-1}\mathbf{A}_h$ was found to be bounded in [10]. Furthermore, the bound was found to be particularly low when GLL nodes and quadrature were used for both $\mathbf{A}_h$ and $\mathbf{A}_\ell$. To this end, we consider approximations of this bound by

Table **1**: Computational results for FOSLS diffusion problem for a 64x64 finite element mesh. NIT refers to the number of PCG iterations to reduce the residual by a factor of 10^{-10} for the high-order system where the AMG preconditioner is either built from the high-order operator (HO) or the sparse optimized low-order operator (LO).

Order	NIT_{HO}	NIT_{LO}	Functional Norm
1st	14	14	2.66e+01
2nd	24	46	1.63e$-$02
3rd	21	62	4.43e$-$06
4th	31	81	6.77e$-$10
5th	33	97	6.59e$-$14

Table **2**: Approximation of $\kappa_2(\mathbf{A}_\ell^{-1}\mathbf{A}_h)$ using CL and GLL points for 2D Poisson's equation on a sheared square domain. DS results are for the discretized sparse operator and LS results are for the least-squares optimized sparse operator.

$\mathbf{A}_\ell$ Type	2nd	4th	6th	8th	10th
DS-GLL	3.1	3.6	4.1	4.5	4.8
LS-GLL	1.6	2.5	2.8	2.9	2.9
DS-CL	3.4	5.5	8.3	10	11
LS-CL	1.8	3.3	3.6	3.9	4.1

looking at condition number estimates for a single element on which we impose pure Neumann BCs. We examine these bounds for the different quadrature schemes and the two low-order preconditioners considered in this work. Since $\mathbf{A}_\ell$ is a singular matrix for a single element, the pseudoinverse of $\mathbf{A}_\ell$ must be used so we form $\mathbf{A}_\ell^+\mathbf{A}_h$. Furthermore, we do not use the smallest singular value of this product in computing the condition number, $\kappa_2(\mathbf{A}) = \sigma_{max}/\sigma_{min}$, but instead we use the smallest nonzero singular value. Singular values were numerically found in Python using the NumPy package. We consider Poisson's equation for GLL (2D only) and CL (2D and 3D) points on sheared square and sheared cube domains. Results are presented in Table **2** and **3**.

Table **3**: Approximation of $\kappa_2(\mathbf{A}_\ell^{-1}\mathbf{A}_h)$ using CL for 3D Poisson's equation on a sheared cube domain. DS results are for the discretized sparse operator and LS results are for the least-squares optimized sparse operator.

$\mathbf{A}_\ell$ Type	2nd	4th	6th
DS-CL	12	21	41
LS-CL	2.7	8.7	10

5.3   Convergence results

For convergence results, we begin with 2D Poisson's equation defined on the sheared square domain. We fix the number of degrees of freedom at 58,081 and examine convergence for 2nd- to 10th-order polynomials, while varying the mesh size to keep the number of degrees of freedom fixed. Results for GLL and CL nodes using the low-order discretization, the least-squares optimized method, and a direct application of AMG on the high-order system are presented in Fig. (**5**). For 3D Poisson's equation on a sheared cube domain, we fix the number of degrees of freedom at $\sim 550,000$ DOFs and examine convergence for 2nd- to 6th-order in Fig. (**6**) for CL nodes. GLL nodes were not an option in the 3D code.

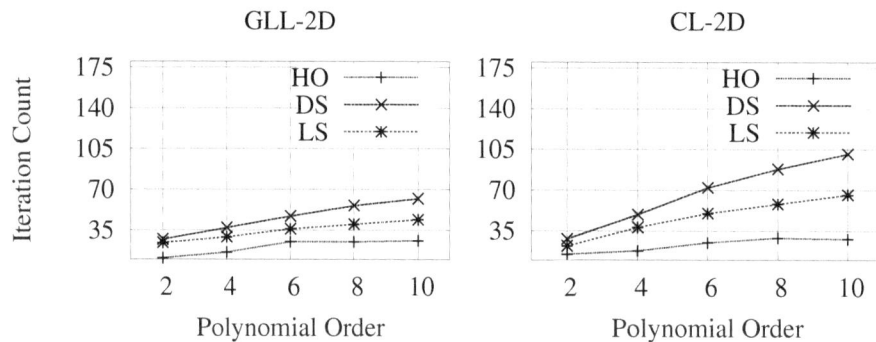

Figure (**5**): (Fixed problem size) Number of iterations required by CG using an AMG preconditioner to reduce the residual by a factor of 10^{-10} for 2D Poisson's equation on a sheared square domain. The problem size is fixed at 58081 DOFs. The left plot corresponds to using GLL points for quadrature points and node locations and the right plot corresponds to using CL for node locations and exact quadrature. HO data is for an AMG preconditioner built from the high-order operator, DS data is for an AMG preconditioner built from the low-order discretized operator, and LS data is for an AMG preconditioner built from the low-order least-squares operator.

In Figs. (**7**)-(**8**), we take the same problem and data but plot the number of Work Units (WUs) required to reduce the residual by a factor of 10^{-10}. A WU is the cost of one residual computation with the high-order matrix for that polynomial order. This allows us to see the required work to solve the problem using the various AMG preconditioners since a single V(1,1)-cycle of AMG built from the high-order matrix costs more than a single V(1,1)-cycle of AMG built from one of the low-order operators. We discuss the conclusions drawn from this data in Section 6.

Next, we analyze 2D and 3D Poisson's equation by fixing the mesh size, instead of fixing the problem size and varying the mesh. Memory results presented along with work estimate results will illustrate the higher polynomial order that we can consider for a given memory amount. In 2D we examined a 32x32 mesh and polynomials ranging from 2nd- to 10th-order. The results are presented in Fig. (**9**) with memory usage in Table **4**. In 3D we examine a 10x10x10 mesh and polynomials ranging from 2nd- to 6th-order. For CL nodes the work results are presented in Fig. (**10**) with memory usage in Table **5**.

Besides Poisson's equation on the sheared square and sheared cube domains, we have also examined the discontinuous steady diffusion equation and the anisotropic steady diffusion equation on unit domains as well as sheared domains. The results for this wider set of problems show similar behavior

CL-3D

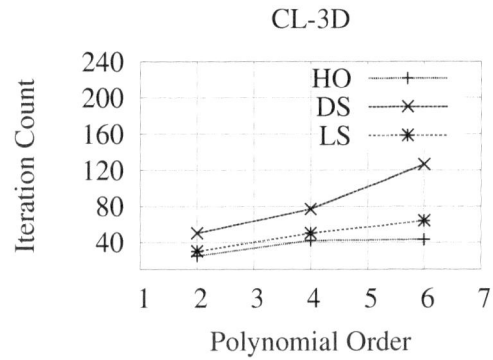

Figure **(6)**: (Fixed problem size) Number of iterations required by CG using an AMG preconditioner to reduce the residual by a factor of 10^{-10} for 3D Poisson's equation on a sheared cube domain. For a given polynomial order, the mesh size is adjusted to fix the problem size at $\sim$ 550,000 DOFs. Data is only generated for CL nodes in 3D since GLL nodes were not at option.

GLL-2D   CL-2D

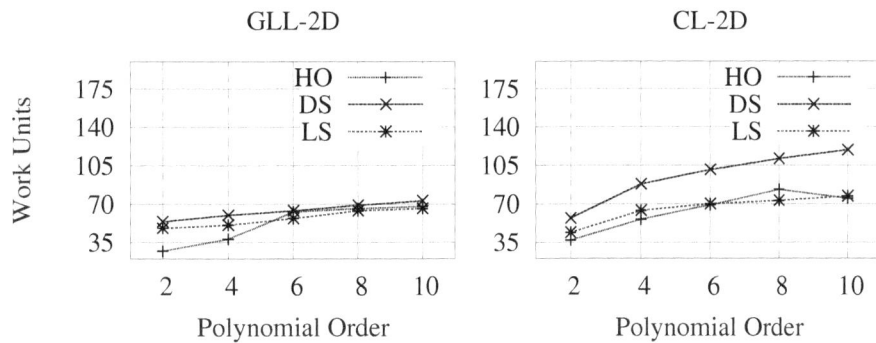

Figure **(7)**: (Fixed problem size) Number of Work Units required by CG using an AMG preconditioner to reduce the residual by a factor of 10^{-10} for 2D Poisson's equation on a sheared square domain. In contrast to Fig. **(5)** examining Work Units allows real comparison of performance of the various approaches.

CL-3D

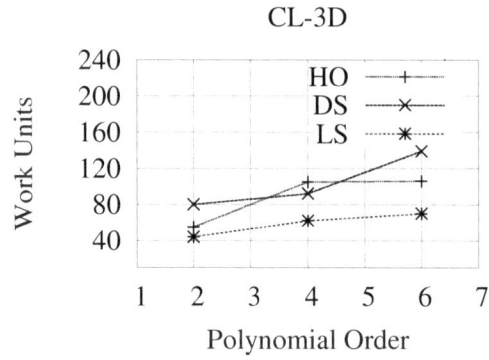

Figure **(8)**: (Fixed problem size) Number of Work Units required by CG using an AMG preconditioner to reduce the residual by a factor of 10^{-10} for 3D Poisson's equation on a sheared cube domain. In contrast to Fig. **(6)** examining Work Units allows real comparison of performance of the various approaches.

GLL-2D   CL-2D

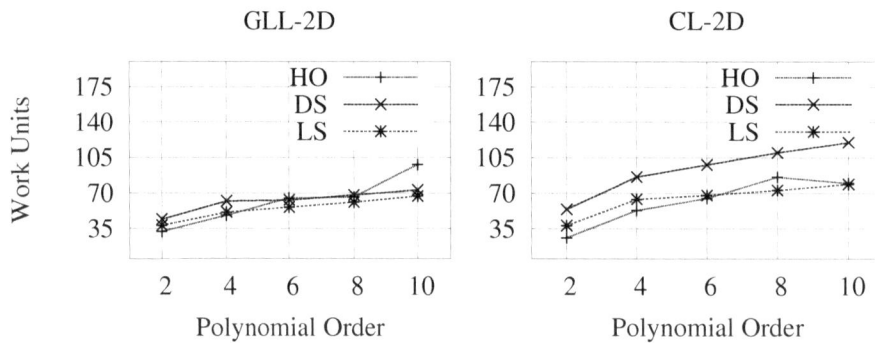

Figure **(9)**: (Fixed mesh size) Number of Work Units required by CG using an AMG preconditioner to reduce the residual by a factor of 10^{-10} for 2D Poisson's equation on a sheared square domain. Various polynomial orders are examined on a 32x32 finite element mesh. See Table **4** for memory usage.

Table **4**: Memory usage in MBs of the high-order matrix (HO-MAT), AMG generated from the high-order matrix (HO-AMG), AMG generated from the low-order discretized matrix (DS-AMG), and AMG generated from the sparse optimized matrix (LS-AMG) for the problems in Fig. **(9)**.

2D System	2nd	4th	6th	8th	10th
HO-MAT-GLL	0.5	4.5	18	50	112
HO-AMG -GLL	0.7	7.4	29	72	183
DS-AMG -GLL	0.5	3.3	6.3	11	20
LS-AMG-GLL	0.5	2.3	5.4	11	20
HO-MAT-CL	0.5	4.5	18	50	112
HO-AMG-CL	0.7	9.4	31	93	187
DS-AMG-CL	0.5	3.6	7.2	13	18
LS-AMG-CL	0.5	3.1	7.1	13	20

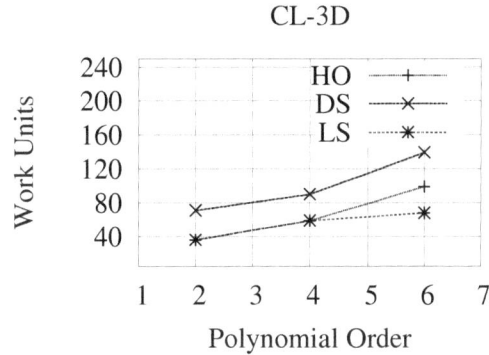

Figure **(10)**: (Fixed mesh size) Number of Work Units required by CG using an AMG preconditioner to reduce the residual by a factor of 10^{-10} for 3D Poisson's equation on a sheared square domain. Various polynomial orders are examined on a 10x10x10 finite element mesh. See Table **4** for memory usage.

to Poisson's equation on sheared square and sheared cube domains. For a fixed polynomial order, we have also investigated convergence behavior as the mesh size is refined and observed numerically that the iteration count is bounded. In the previous section, we showed our ability to solve a diffusion equation, posed using a FOSLS formulation, with an inhomogeneous source from a prescribed analytic solution.

6 Discussion

6.1 Performance comparison

In Figs. **(5)**-**(6)**, we observe a growth in iteration count with polynomial order. As the order increases, so does the condition number of the linear system leading to the growth in iteration count. When employing non-uniform GLL or CL nodes as in Fig. **(1)**, the linear system also becomes more ill-

Table **5**: Memory usage in MBs of the high-order matrix (HO-MAT), AMG generated from the high-order matrix (HO-AMG), AMG generated from the low-order discretized matrix (DS-AMG), and AMG generated from the sparse optimized matrix (LS-AMG) for the problems in Fig. **(10)**.

3D System	2nd	4th	6th
HO-MAT	3.9	106	849
HO-AMG	4.7	160	1250
DS-AMG	2.4	20	92
LS-AMG	2.0	26	85

conditioned due to the mesh anisotropy. The mesh anisotropy also impacts the condition number of the low-order systems as we are effectively discretizing a finite element problem on a non-uniform mesh having elements with widely varying aspect ratios. Whether using an AMG method built from the high-order system or either of the two low-order systems, the effectiveness of the single AMG V-cycle is going to be diminished because of this anisotropy. These same conclusions apply to 3D as well.

Tables **2-3** suggest the optimized least-squares approach should yield better performance than the discretized approach, particularly when using CL nodes. According to Figs. **(5)**-**(6)**, the solution strategy using an AMG preconditioner built from the high-order system is the most effective in terms of iteration counts, but Figs. **(7)**-**(8)** tell a different story when the actual work required to find the solution is considered. Fig. **(7)** shows that the sparse least-squares approach is slightly more effective beyond 6th-order polynomials in 2D. For example, for 8th-order polynomials, the HO approach required 83 WUs for CL nodes and the LS approach required 73 WUs for CL nodes. Similarly, for GLL nodes, the HO approach required 66 WUs and the LS approach required 63 WUs. Additionally, the LS approach requires only one-seventh the memory to store the AMG hierarchy. Fig **(8)** shows that the optimized least-squares approach is consistently more effective beyond 2nd-order polynomials in 3D.

In Figs. **(9)**-**(10)**, for a fixed mesh size, we see similar behavior as the polynomial order is increased. The dimension of the matrix grows, which negatively impacts the condition number of the system and the iteration counts, and the matrix also becomes denser. The most telling result is the memory consumption found in Table **4**, which shows the memory consumed by the original high-order matrix (HO-MAT) and the memory consumed by the AMG hierarchy based on the high-order matrix (HO-AMG), the discretized sparse system (DS-AMG), and the optimized sparse system (LS-AMG). We can see that if we only have to store the AMG hierarchy then we can use 10th-order polynomials with the sparse AMG system at the same memory cost of 6th-order polynomials with the high-order system. We can expect such a memory savings when a matrix-free implementation of the high-order system is used, such as in [19].

For all examples, we held the number of V-cycles fixed in the preconditioning stage and also fixed the strength of connection threshold from the AMG algorithm. For 2D problems we increased the number of V-cycles allowed and saw little reduction in the total number of PCG iterations required to reduce the tolerance to the desired level. In the same vein, we tested varying the strength of connection threshold (see [30]) and again saw no significant change in the number of iterations required to reduce the tolerance to the desired level. No further tests were performed to find better AMG performance based on a different set of AMG parameters.

6.2   Cost comparison

In Figs. **(7)**-**(8)**, we see the actual cost in terms of WUs to perform the solve given a preconditioner type. However, this is only a part of the total cost to use the various preconditioners. When using AMG as a preconditioner, there is also the cost of setting up the entire AMG hierarchy. This cost depends on stencil size as well as a variety of tuning parameters from the AMG algorithm. In addressing setup costs, we simply select the default AMG parameters and compare the cost of setting up the AMG hierarchy. As for the sparse systems, we only consider the AMG setup costs of the optimized least-squares version described in Section 4. For 8th-order finite elements on a 30×30 mesh, the AMG hierarchy for the high-order system required 15 s for the setup while the AMG hierarchy for the sparse system required only 0.87 s for the setup. We clearly get a savings in AMG setup as well as savings in memory and savings in solution time.

Next, we consider the cost of setting up the sparse preconditioners. For the discretized approach the cost is equal to the cost of discretizing the problem with low-order finite elements. We do not provide timings here but note that this also requires the user to provide a new code that performs this discretization. We aim to avoid asking the user to perform this work, but instead merely provide access to the high-order element stiffness matrices. Thus, we address now the cost of setting up the sparse matrix with the optimized least-squares approach of Section 4. We have focused in this work on simple uniform diffusion problems where both the mesh and the coefficients are constant. For this approach to be viable, we must consider it in the context of varying mesh and/or coefficients.

To construct a sparse representation of a high-order *dense* element stiffness matrix, we must setup and solve a least-squares problem. First, we have to solve an eigenvalue problem for $\mathbf{A}_e^H$. In N dimensions for pth-order finite elements, $\mathbf{A}_e^H$ is a $(p+1)^N \times (p+1)^N$ dense matrix. The number of flops to find the eigenspace of this system is $O((p+1)^{3*N})$. In 2D, for 8th-order polynomials, the time to find the eigenspace using LAPACK is 0.02 s for a single element. On a 30×30 mesh, the time to setup the AMG hierarchy for the sparse system is 0.87 s, while the time to setup the AMG hierarchy for the high-order system was 15 s. To find the eigenspace for 900 elements the time to find the eigenspace is going to be 18 s. Since this is a very local problem, we believe that we can reduce this time by taking advantage of multicore architectures. Furthermore, we are exploring means of constructing the system defined by Eq. (4.1) that does not require finding the whole eigenspace.

The next stage of the setup is solving the linear system associated with Eq. (4.1). We are currently using an inefficient dense LAPACK solve. For the same problem as discussed above the time to solve this system for a single element was 0.016 s. Over all 900 elements the time to solve all systems was 14 s. We are currently exploring a variety of approaches to quickly solving this linear system from iterative approaches to approaches that take advantage of the block structure of the system. Preliminary results suggest that to reduce the overall setup of the sparse element stiffness matrix we should focus on reducing the cost of finding the eigenspace since better solvers than LAPACK can reduce the time to solve the linear system that defines the sparse stiffness matrix. We present these results in greater detail in the future.

Finally, we note that even if we can reduce the cost of a single element stiffness matrix setup to the minimal value, we still should avoid setting up this sparse element stiffness matrix for every single element. For varying mesh or coefficients, we will have to reduce the need to setup the sparse element stiffness matrix at each element and instead use a single representative sparse element stiffness matrix as an approximation. Such approaches have been used before in the work of Fischer *et al.* in [31]. An example of this is for a slowly varying mesh with near constant elemental area or volume, we can setup one sparse element stiffness matrix and use this throughout as an approximation. This will be explored in a future paper.

7 Conclusions

We have presented a means of solving a high-order finite element problem with a low-order system that does not require a rediscretization of the problem. Instead, a small eigenvalue problem in combination with a least-squares solve is used. We have shown this new approach delivers better performance and the same reduction in memory as a rediscretization. The better performance comes at the cost of solving the local eigenvalue problem and the least-squares problem. We are currently investigating the means of overcoming the eigenvalue solve and reducing the cost of the least-squares solve. We can examine, for example, new modern multicore processors that allow large amounts of local computation and have been included in many of the new, exascale computing architectures. We are also addressing variable coefficients or nonuniform meshes that require updating the sparse element stiffness matrix for each element.

Bibliography

[1] Axelsson O, Gustafsson I. Preconditioning and two-level multigrid methods of arbitrary degree of approximation. Math Comp 1983;40(161):219–242.

[2] Maitre J, Musy F. The contraction number of a class of two-level methods: an exact evaluation for some finite element subspaces and model problems. In: Hackbusch W, Trottenberg U, editors. Multigrid Methods, eds., Lecture Notes in Mathematics. vol. 960. Springer-Verlag; 1982. pp. 535–554.

[3] Mandel J. On block diagonal and Schur complement preconditioning. Numer Math 1990;58(1):79–93.

[4] Deville MO, Fischer PF, Mund EH. High-Order Methods for Incompressible Fluid Flow. Cambridge University Press; 2002.

[5] Mandel J. Two-level domain decomposition preconditioning for the *p*-version finite element method in three dimensions. Int J Numer Methods Eng 1990;29(5):1095–1108.

[6] Pavarino LF, Widlund OB. A polylogarithmic bound for an iterative substructuring method for spectral element in three dimensions. SIAM J Numer Anal 1996;33(4):1303–1335.

[7] Casarin MA. Diagonal edge preconditioners in p-version and spectral element methods. SIAM J Sci Comput 1997;18(2):610–620.

[8] Casarin MA. Quasi-optimal Schwarz methods for the conforming spectral element discretization. SIAM J Numer Anal 1997;34(6):2482–2502.

[9] Dryja M, Smith BF, Widlund OB. Schwarz analysis of iterative substructuring algorithms for elliptic problems in three dimensions. SIAM J Numer Anal 1994;31(6):1662–1694.

[10] Fischer PF. An Overlapping Schwarz Method for Spectral Element Solution of the Incompressible Navier-Stokes Equations. J Comp Phys 1997;133(1):84–101.

[11] Fischer PF, Ronquist EM. Spectral element methods for large-scale parallel Navier-Stokes calculations. Comput Method Appl M. 1994;116(1–4):69–76.

[12] Tufo HM, Fischer PF. Fast Parallel Direct Solvers for Coarse Grid Problems. J Parallel Distr Com 2001;61:151–177.

[13] Lottes JW, Fischer PF. Hybrid Multigrid/Schwarz Algorithms for the Spectral Element Method. J Sci Comp 2005;24(1):45–78.

[14] Orszag SA. Spectral methods for problems in complex geometries. J Comp Phys 1980;37(1):70–92.

[15] Deville M, Mund E. Cheybshev pseudospectral solution of second-order elliptic equations with finite element preconditioning. J Comp Phys 1985;60(3):517–522.

[16] Shen J, Wang F, Xu J. A finite element multigrid preconditioner for Chebyshev-collocation methods. App Num Math 2000;33(1):471–477.

[17] Kim SD, Parter SV. Preconditioning Chebyshev spectral collocation by finite-difference operators. SIAM J Numer Anal 1997;34(3):939–958.

[18] Heys JJ, Manteuffel TA, McCormick SF, Olson LN. Algebraic multigrid for higher-order finite elements. J Comp Phys 2005;204(2):520–532.

[19] Brown J. Efficient Nonlinear Solvers for Nodal High-Order Finite Elements in 3D. J Sci Comp 2010;45(1):48–63.

[20] Brezina M, Cleary AJ, Falgout RD, Henson VE, Jones JE, Manteuffel TA, et al. Algebraic multigrid based on element interpolation (AMGe). SIAM J Sci Comput 2001;22(5):1570–1592.

[21] Carey G, Barragy E. Basis function selection and preconditioning high degree finite element and spectral methods. BIT 1989;29(4):794–804.

[22] Šolín P, Segeth K, Doležel I. Higher-order finite element methods, Studies in Advanced Mathematics. Chapman & Hall/CRC; 2003.

[23] Ruge JW. AMG for Higher-Order Discretizations of Second-Order Elliptic Problems. In: Abstracts of the Eleventh Copper Mountain Conference on Multigrid Methods; 2000.

[24] Durufle M, Grob P, Joly P, Rocquencourt F. Influence of Gauss and Gauss-Lobatto quadrature rules on the accuracy of a quadrilateral finite element method in the time domain. Numer Methods Partial Differential Equations. 2009;25(3):526–551.

[25] Brenner SC, Scott LR. The mathematical theory of finite element methods. Springer Verlag; 2002.

[26] Parter SV. Preconditioning Legendre spectral collocation methods for elliptic problems II: Finite element operators. SIAM J Numer Anal 2002;39(1):348–362.

[27] Olson L. Algebraic multigrid preconditioning of high-order spectral elements for elliptic problems on a simplicial mesh. SIAM J Sci Comput 2007;29(5):2189–2209.

[28] Cai Z, Manteuffel TA, McCormick SF. First-order system least squares for second-order partial differential equations: Part II. SIAM J Numer Anal 1997;34(2):425–454.

[29] Henson VE, Yang UM. BoomerAMG: A parallel algebraic multigrid solver and preconditioner. App Num Math 2002;41(1):155–177.

[30] Briggs WL, Henson VE, McCormick SF. A Multigrid Tutorial: Second Edition. 2nd ed. Pennsylvania, PA: SIAM; 2001.

[31] Fischer PF, Miller NI, Tufo HM. An overlapping Schwarz method for spectral element simulation of three-dimensional incompressible flows. Parallel Solution of Partial Difierential Equations, Springer Verlag, Berlin. 2000; pp. 159–181.

Chapter 6

A Geometric Toolbox for Tetrahedral Finite Elements Partitions

Jan Brandts[1], Sergey Korotov[2], and Michal Křížek[3]

Abstract: In this work we present a survey of some geometric results on tetrahedral partitions and their refinements in a unified manner. They can be used for mesh generation and adaptivity in practical calculations by the finite element method (FEM), and also in theoretical finite element (FE) analysis. Special emphasis is laid on the correspondence between relevant results and terminology used in FE computations, and those established in the area of discrete and computational geometry (DCG).

Keywords: *finite element method, tetrahedron, polyhedral domain, linear finite element, angle and ball conditions, convergence rate, mesh regularity, discrete maximum principle, mesh adaptivity, red, green and yellow refinements, bisection algorithm*

1 Introduction and Motivation

Many geometric facts about tetrahedra and partitions of polyhedra into tetrahedra are known, and some of them already for quite some time. Even so, with the appearance and permanent growth in speed and capacity of modern computers, together with the practical needs originating from various numerical methods such as the finite element method (FEM), new challenges still appear in this context.

Tetrahedra seem to be the most natural "basic shapes" for dissection or approximation of complicated 3D domains. As a result, constructing tetrahedral partitions and their refinements are among the most challenging problems in finite element discretization of three-dimensional partial differential equations that arise for instance in mathematical physics and engineering. In this survey, we discuss both mathematical and numerical issues related to this topic.

To start, we briefly present two motivating examples. First, it is commonly believed, not only among FEM practitioners but also in discrete and computational geometry (DCG), that the use of near degenerate tetrahedra in a partition should, if possible, be avoided. However, we will point out (see

[1] Korteweg-de Vries Institute, University of Amsterdam, Netherlands; e-mail: janbrandts@gmail.com

[2] (corresponding author) Basque Center for Applied Mathematics, Bizkaia Technology Park, Derio, Basque Country, Spain; e-mail: korotov@bcamath.org

[3] Institute of Mathematics, Academy of Sciences, Prague, Czech Republic; e-mail: krizek@math.cas.cz

Owe Axelsson and János Karátson (Eds)

Section 4) that not all such tetrahedra are that bad, and, moreover, that some are even unavoidable in certain situations, e.g. for covering thin slots, gaps or strips of different materials (see [1, p. 76]). Second, note that a single obtuse triangle or tetrahedron in a finite element triangulation can destroy the validity of the discrete maximum principle (DMP) for the Poisson equation $-\Delta u = f$ with homogeneous Dirichlet boundary conditions (see e.g. [2]). For instance, let the domain $(0,4) \times (0,2)$ be triangulated as in Figure **(1)** below. The space of continuous piecewise linear functions relative to this triangulation that satisfy the boundary conditions has dimension three. Their degrees of freedom are the values at the vertices $v_1 = (1,1), v_2 = (3,1)$, and $v_3 = (2,1+p)$, which are indicated with dots. The triangle with vertices v_1, v_2, v_3 is obtuse for all $p \in (0,1)$. It can be easily verified that the discrete Laplacian does not have a non-negative inverse. For example, for $p = \frac{1}{2}$ this inverse equals

$$\begin{pmatrix} \frac{63}{248} & -\frac{1}{248} & \frac{1}{16} \\\\ -\frac{1}{248} & \frac{63}{248} & \frac{1}{16} \\\\ \frac{1}{16} & \frac{1}{16} & \frac{37}{160} \end{pmatrix}.$$

Therefore, each non-positive continuous function $f \neq 0$ whose support does not intersect the supports of the finite element functions that vanish at v_1, gives rise to an approximation u_h of u that is positive at v_2, hence violating the DMP (cf. Remark 6.11 below).

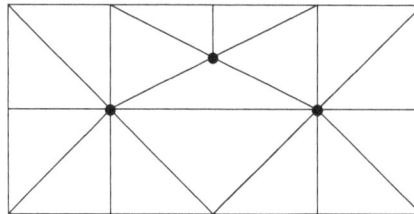

Figure **(1)**: Triangulation with a single obtuse triangle for $p = \frac{1}{2}$.

The above two examples show that the geometric properties of partitions used are important. In what follows, we assume that $\Omega \subset \mathbf{R}^3$ is a given domain. If the boundary $\partial\overline{\Omega}$ of $\overline{\Omega}$ is contained in a finite number of planes, then $\overline{\Omega}$ is called a *polyhedral domain*. If $\overline{\Omega}$ is bounded, it is called a *polyhedron*. Further, let $L^2(\Omega)$ be the space of square integrable functions over Ω equipped with the standard norm. Sobolev spaces are denoted by $H^s(\Omega)$. The symbol c stands for a generic constant, and vol_d stands for the d-dimensional Euclidean volume.

2 Tetrahedra

2.1 Main geometric characteristics

Let $A = (A_1, A_2, A_3)$, $B = (B_1, B_2, B_3)$, $C = (C_1, C_2, C_3)$, and $D = (D_1, D_2, D_3)$ be points in $\mathbb{R}^3$ that are not contained in one plane. We denote by T the tetrahedron with vertices A, B, C, and D (see Figure **(2)**). It is the simplest closed convex polyhedron, which has 4 triangular faces and 6 edges. The volume of T can, for example, be calculated by the following formula:

$$\text{vol}_3 T = \frac{|\delta|}{6}, \tag{2.1}$$

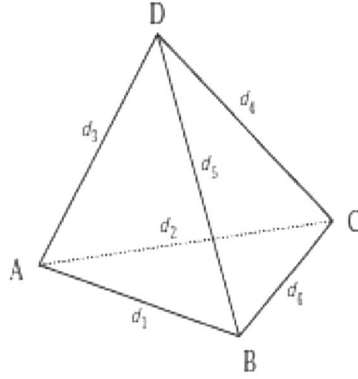

Figure **(2)**: Tetrahedron T with denotation.

where

$$\delta = \det \begin{bmatrix} B_1 - A_1 & B_2 - A_2 & B_3 - A_3 \\ C_1 - A_1 & C_2 - A_2 & C_3 - A_3 \\ D_1 - A_1 & D_2 - A_2 & D_3 - A_3 \end{bmatrix} = \det \begin{bmatrix} 1 & A_1 & A_2 & A_3 \\ 1 & B_1 & B_2 & B_3 \\ 1 & C_1 & C_2 & C_3 \\ 1 & D_1 & D_2 & D_3 \end{bmatrix}, \tag{2.2}$$

see [3, Sect. 6.2]. Further,

$$r_T = \frac{3 \operatorname{vol}_3 T}{\operatorname{vol}_2 \partial T} \tag{2.3}$$

is the radius of the inscribed ball of T, where ∂T is the boundary of T.
By [4, p. 316], the radius of the circumscribed ball about T can be computed as

$$R_T = \frac{\sqrt{Z_T}}{24 \operatorname{vol}_3 T}, \tag{2.4}$$

where

$$Z_T = 2d_1^2 d_2^2 d_4^2 d_5^2 + 2d_1^2 d_3^2 d_4^2 d_6^2 + 2d_2^2 d_3^2 d_5^2 d_6^2 - d_1^4 d_4^4 - d_2^4 d_5^4 - d_3^4 d_6^4. \tag{2.5}$$

In the above formula d_i and d_{i+3} are the Euclidean lengths of opposite edges of T for $i = 1, 2, 3$:

$$d_1 = \|A - B\|, \ d_2 = \|A - C\|, \ d_3 = \|A - D\|, \ d_4 = \|C - D\|, \ d_5 = \|B - D\|, \ d_6 = \|B - C\|.$$

See Figure **(2)**.
The dihedral angles of a tetrahedron are the six angles between each pair of faces of T. They are defined as the complementary angles of outward unit normals to those facets and can be calculated by means of the inner product (see [5, p. 385], [6]):

$$\cos \alpha = -n_1 \cdot n_2, \tag{2.6}$$

where n_1 and n_2 are outward unit normals of particular faces.

2.2　On the shapes of tetrahedra

It is common in both FE analysis and DCG to qualitatively distinguish between so-called "well-shaped" (i.e. close to regular) tetrahedra and "badly-shaped" ones (i.e. close to degenerate). Some classifications of badly-shaped tetrahedra are given in [7, p. 191], [8, p. 3], [9, p. 195], [10, p. 286], and [11, p. 256].

The classification in Figure (**3**) is taken from [8, 9], which distinguishes between so-called "skinny" and "flat" badly-shaped tetrahedra, based on closeness of vertices of such tetrahedra to one line (skinny) or to one face (flat). In practice (see [12, p. 794]), the degree of degeneration of a tetra-

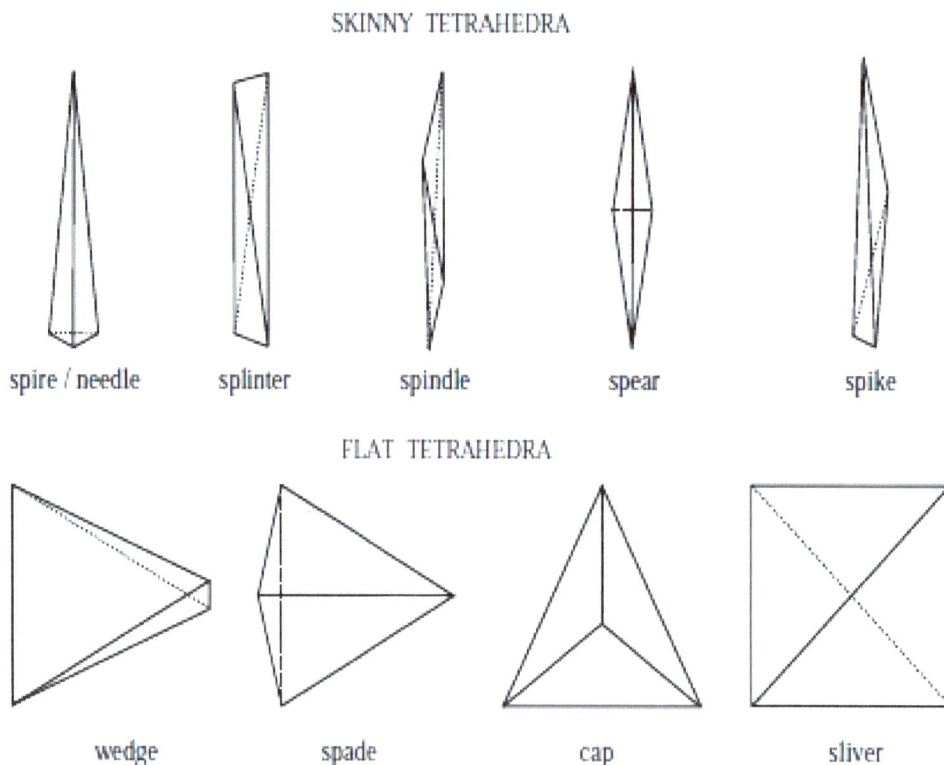

SKINNY TETRAHEDRA

spire / needle　　　splinter　　　spindle　　　spear　　　spike

FLAT TETRAHEDRA

wedge　　　spade　　　cap　　　sliver

Figure (**3**): Classification of "badly-shaped" tetrahedra according to [8, 9]. However, some tetrahedra (needles, splinters, wedges) satisfy the maximum angle condition (see (4.5)–(4.6) in below), which guarantees that their shape does not influence the nodal interpolation error (and, therefore, theoretical and also practical convergence of FE approximations) in a negative way.

hedron T is often measured in terms of the *quality indicator*

$$Q_T = 3\frac{r_T}{R_T} \in (0,1],\tag{2.7}$$

with r_T and R_T defined in (2.3) and (2.4). Tetrahedra with quality indicator Q_T near 1 are almost regular, whereas those with Q_T near 0 are nearly degenerate. Other quality indicators used in DCG and FEMs (and their comparison) can be found e.g. in [9–13].

In Section 4 we shall introduce several regularity conditions in terms of angles and balls that are used in finite element convergence proofs.

3   Tetrahedral Partitions of Polyhedral Domains

3.1   On face-to-face partitions of polyhedra into tetrahedra

Definition 6.1. A finite set of tetrahedra is a (face-to-face) partition of a polyhedron $\overline{\Omega}$ if

i) the union of all the tetrahedra is $\overline{\Omega}$,

ii) the interiors of the tetrahedra are mutually disjoint,

iii) any face of any tetrahedron from the set is either a face of another tetrahedron in the set, or a subset of $\partial\Omega$.

Alternative terminology (commonly used in both FEM and DCG) is a simplicial complex, decomposition, dissection, division, grid, lattice, mesh, net, network, triangulation, space discretization, subdivision, tetrahedralization, etcetera.

Theorem 6.1. *For any polyhedron there exists a partition into tetrahedra.*

The main idea of the detailed constructive proof presented in [1,14] is the following. Denote the faces of a given polyhedron $\overline{\Omega}$ by $F_1, \ldots, F_m$. Consider the planes $P_1, \ldots, P_m \subset \mathbb{R}^3$ such that

$$F_i \subset P_i, \ i = 1, \ldots, m.$$

It can be shown that all components of the set

$$\overline{\Omega} \setminus \cup_{i=1}^{m} P_i$$

are open convex polyhedra. Their closures can be decomposed into tetrahedra as follows. First, we triangulate each of its polygonal faces as sketched in Figure (**4**). Second, we take the convex hull of the gravity center of the convex polyhedron with each of the triangles on its surface. If all common faces are triangulated in the same way, a partition of $\overline{\Omega}$ into tetrahedra satisfying the conditions of Definition 6.1 results.

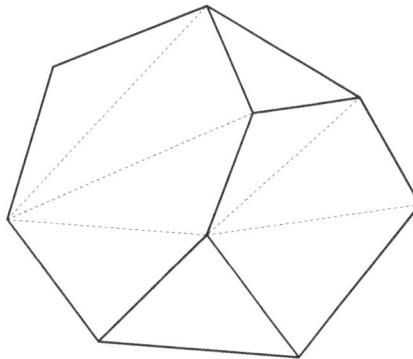

Figure (**4**): Partition of a convex polyhedron into tetrahedra. Each polygonal face of its surface is divided into triangles.

For a given partition $\mathcal{T}_h$ the *discretization parameter* h stands for the maximum length of all edges in the partition, i.e.,

$$h = \max_{T \in \mathcal{T}_h} h_T,$$

where

$$h_T = \operatorname{diam} T.$$

3.2   Various refinement techniques for tetrahedral partitions

In FE analysis and computation, one needs sequences (infinite or finite) of partitions that have certain properties. They are usually constructed by face-to-face refinements of a given coarse partition [15, 16].

Definition 6.2. An infinite sequence $\mathcal{F} = \{\mathcal{T}_h\}_{h\to 0}$ of partitions of $\overline{\Omega}$ is called a *family of partitions* if for every $\varepsilon > 0$ there exists $\mathcal{T}_h \in \mathcal{F}$ with $h < \varepsilon$.

One can define various kinds of "well-shapedness", usually called regularity, in the sense that certain properties of the tetrahedral elements are supposed to hold uniformly over all partitions of the family.

Definition 6.3. A family $\mathcal{F}$ of partitions is *regular (strongly regular)* if there exists a constant $c > 0$ such that for any $\mathcal{T}_h \in \mathcal{F}$ and any $T \in \mathcal{T}_h$ we have

$$\text{vol}_3 T \geq c h_T^3 \quad (\text{vol}_3 T \geq c h^3). \tag{3.1}$$

Remark 6.1. It is easy to construct strongly regular families of triangulations of a polygonal domain into triangles in the sense that $\text{vol}_2 T \geq c h^2$. This is because each triangle can be subdivided into four congruent triangles similar to the original one. Also techniques based on bisection can be used, see e.g. [17] for details. In three dimensions it is generally not possible to subdivide a tetrahedron into congruent tetrahedra similar to the original one (cf. [14]).

Nevertheless, the following theorem is valid.

Theorem 6.2. *For any tetrahedron there exists a strongly regular family of partitions into tetrahedra.*

For a detailed constructive proof see [14], or [1]. The main idea is that the reference tetrahedron $\tilde{T} = \tilde{A}\tilde{B}\tilde{C}\tilde{D}$, whose opposite edges $\tilde{A}\tilde{B}$ and $\tilde{C}\tilde{D}$ have length 2 and the length of the remaining edges is $\sqrt{3}$, can be divided into 8 congruent subtetrahedra which are similar to $\tilde{T}$ (cf. Figure (**5**)). This is the only tetrahedron (up to scaling) with such a property. An arbitrary tetrahedron T can now be decomposed into 8 tetrahedra (cf. Figure (**5**)) using an affine one-to-one mapping between $\tilde{T}$ and T. Such a refinement is called *red*.

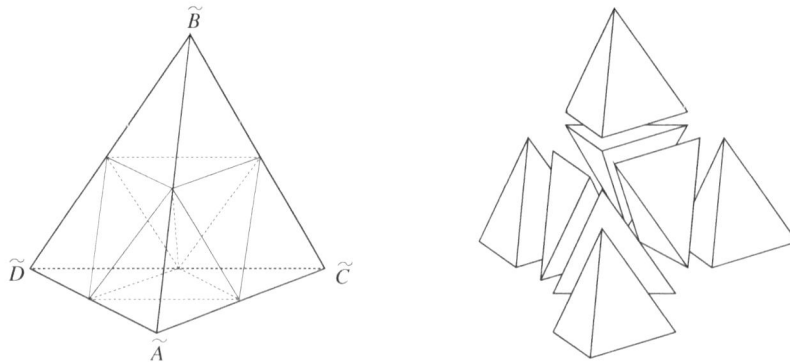

Figure (**5**): Red refinement in 3D.

Remark 6.2. An interesting observation on the performance of the 3D red refinement is presented in [14] and [18]. The convex hull of a vertex of a tetrahedron T with the midpoints of the outgoing edges is a tetrahedron similar to the original one. The octahedron that remains after cutting away the four tetrahedra corresponding to each of the four vertices of T has three spatial diagonals (see Figure (**5**)). Therefore, there are three possibilities for refining a given tetrahedron into 8 subtetrahedra so that its boundary triangles are divided by midlines. However, only choosing the shortest interior diagonal of the octahedron leads to a regular family of tetrahedral face-to-face partitions.

Theorem 6.3. *For any polyhedron there exists a strongly regular family of partitions into tetrahedra.*

Its proof follows immediately from Theorems 6.1 and 6.2.

Local refinements of tetrahedral partitions are needed at those regions in Ω, where singularities or large variations of the solution of PDEs and its derivatives occur. This usually happens near vertices and edges of the polyhedron Ω, or where jumps in coefficients occur, or where the type of boundary condition changes, or near the so-called interfaces (see, e.g., [19, 20]).

In two dimensions, such refinements are usually done with the help of midlines and medians of triangles. Triangles that are divided by midlines are called *red* and by medians *green*, see [21]. The corresponding refinements are also called red and green [22]. Other refinement techniques exist, such as *red** refinement [23], *blue* refinement [24], and *yellow* refinement [25] (see Figure (**6**)).

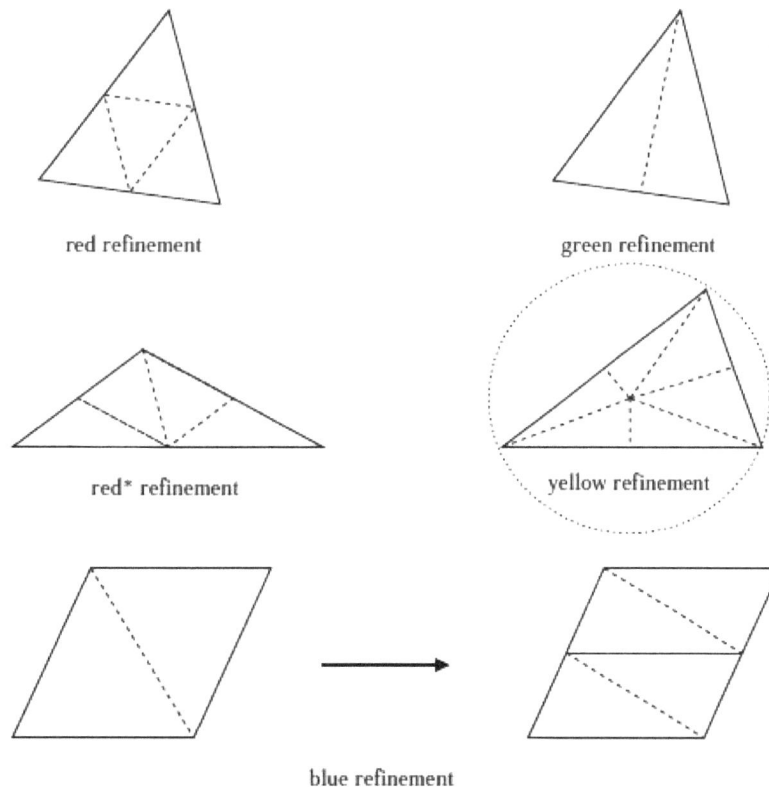

Figure (**6**): Refinement techniques in 2D.

Three-dimensional analogues of green and red refinement are sketched in Figures **(7)** and **(5)**. In Figure **(7)**, we also depict a hybrid red-green refinement: one face of the tetrahedron is divided by midlines and the other faces by medians. A three-dimensional analogue of yellow refinement from Figure **(6)** will be introduced in Section 5.1.

It is worth noting that green refinements from Figures **(6)** and **(7)** are also known in the literature as *bisections*, see e.g. [17, 22, 26].

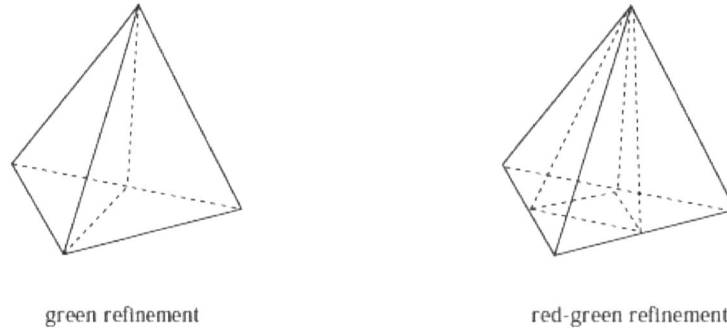

green refinement   red-green refinement

Figure **(7)**: 3D analogues of green and red refinements.

Remark 6.3. In [26], a simple algorithm is presented that generates local refinements of tetrahedral partitions using green and red-green refinement of tetrahedra. They induce a regular family $\mathcal{F}$. Moreover, it can proved that there exists a constant $c > 0$ such that $Q_T \geq c$ for all tetrahedra $T \in \mathcal{T}_h$ and all $\mathcal{T}_h \in \mathcal{F}$, where Q_T is the quality indicator of T defined in (2.7).

Remark 6.4. In Figure **(8)**, we depict another local refinement procedure to treat vertex singularities proposed by B. Guo in [27]. A tetrahedron is first decomposed into one tetrahedron and several pentahedra as in Figure **(8)**a. Then, each pentahedron is decomposed into three tetrahedra as in Figure **(8)**b.

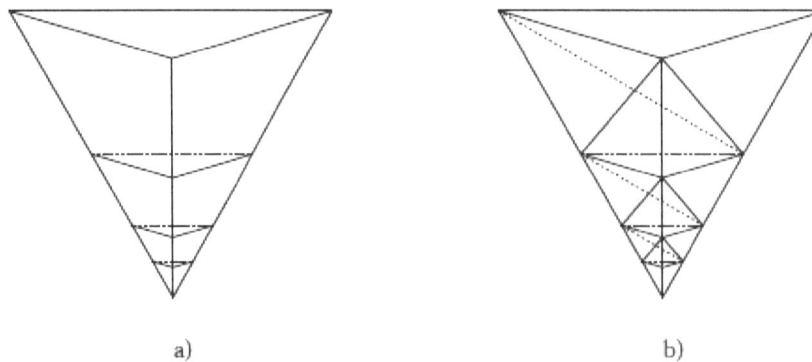

a)   b)

Figure **(8)**: Local refinement technique from [27].

In [5] we describe an algorithm that generates local refinements of nonobtuse tetrahedra towards a vertex. The main idea is based on recursive use of Coxeter's trisection [28] on the left of Figure **(9)**.

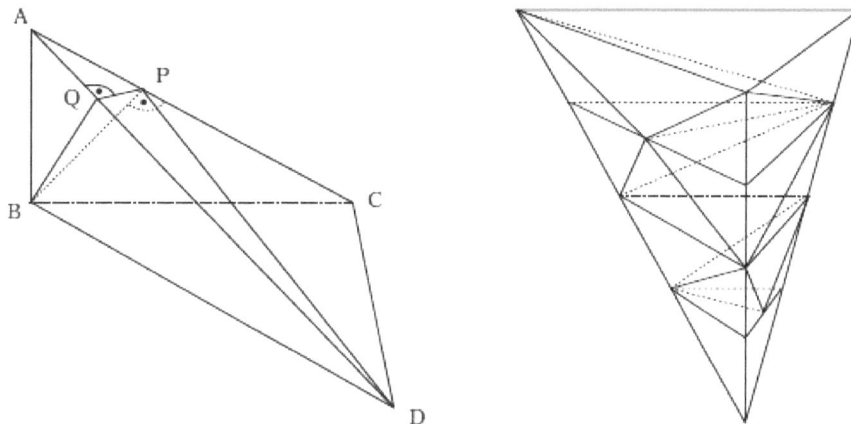

Figure **(9)**: Examples of local nonobtuse refinements towards a vertex.

Remark 6.5. Domains with curved boundaries are usually approximated by polyhedra. Doing this, a so-called variational crime is committed. The remainder of the domain can be handled, e.g. by special curved hat and slice elements. See [29].

Remark 6.6. Uniform or almost uniform partitions usually produce various superconvergence phenomena, see e.g. [30, 31] and references therein.

4   Mesh Regularity: Angle and Ball Conditions in FE Analysis

In [32], one can find results on interpolation estimates for piecewise polynomial functions relative to a family of partitions of the domain, and their relation to the approximation error in FEM. Some of them also follow from the theorems given in the section below.

4.1   The minimum angle condition

The following four regularity conditions for families of simplicial partitions are commonly used in the FE analysis (cf. Section 3.2). The constants c_i in those conditions may depend on the dimension $d \in \{2, 3\}$.

Condition 1: There exists $c_1 > 0$ such that for any $\mathcal{T}_h \in \mathcal{F}$ and any $T \in \mathcal{T}_h$

$$\operatorname{vol}_d T \geq c_1 h_T^d. \tag{4.1}$$

Condition 2: There exists $c_2 > 0$ such that for any $\mathcal{T}_h \in \mathcal{F}$ and any $T \in \mathcal{T}_h$ there exists a ball $b \subset T$ with radius r_T such that

$$r_T \geq c_2 h_T. \tag{4.2}$$

Condition 3: There exists $c_3 > 0$ such that for any $\mathcal{T}_h \in \mathcal{F}$ and any $T \in \mathcal{T}_h$

$$\operatorname{vol}_d T \geq c_3 \operatorname{vol}_d B, \tag{4.3}$$

where $B \supset T$ is the circumscribed ball about T.

Condition 4: There exists $c_4 > 0$ such that for any $\mathcal{T}_h \in \mathcal{F}$, any $T \in \mathcal{T}_h$, and any dihedral angle α and, for $d = 3$, also any angle α within a triangular face of T, we have

$$\alpha \geq c_4. \tag{4.4}$$

Theorem 6.4. *The above four regularity conditions are equivalent for $d = 2, 3$.*

The proof can be found in [33]. Condition 2 is sometimes called the *inscribed ball condition* [32]. Condition 4 is usually called the *minimum angle condition*. In the 2D case it was introduced by M. Zlámal in [34].

Remark 6.7. If the quality factor (2.7) is bounded from below by a constant $c > 0$ independently of h, then the family $\mathcal{F}$ of simplicial partitions is regular, since (4.2) is valid:

$$r_T \geq \frac{c}{3} R_T \geq \frac{c}{6} h_T$$

as $2R_T \geq h_T$.

4.2 Maximum angle condition

Definition 6.4. A family $\mathcal{F}$ of partitions of a polyhedron into tetrahedra is said to be *semiregular* if there exist a $c_5 < \pi$ such that for any $\mathcal{T}_h \in \mathcal{F}$, any $T \in \mathcal{T}_h$, any dihedral angle γ between faces of T and any angle φ within a triangular face of T, we have

$$\gamma \leq c_5, \tag{4.5}$$

$$\varphi \leq c_5. \tag{4.6}$$

The maximum angle condition (4.6) for triangles was first introduced by J. L. Synge [35] and for tetrahedra first by M. Křížek [36].

Theorem 6.5. *Any regular family of partitions of a polyhedron into tetrahedra is semiregular.*

For the proof see [36], where it is also shown that the converse implication does not hold. Semiregular families can contain needles, wedges, and splinters of arbitrary thinness. See Figure **(3)**.
For any tetrahedron T and function $v \in C(T)$, we write $\pi_T v$ for the nodal Lagrange linear interpolant of v on T, further, $\|\cdot\|_{k,\infty,T}$ is the norm and $|\cdot|_{k,\infty,T}$ is the seminorm in the Sobolev space $W^{k,\infty}(T)$.

Theorem 6.6. *Let $\mathcal{F}$ be a semiregular family of partitions of a polyhedron into tetrahedra. Then there exists $c_6 > 0$ such that for any $\mathcal{T}_h \in \mathcal{F}$ and any $T \in \mathcal{T}_h$ we have*

$$\|v - \pi_T v\|_{1,\infty,T} \leq c_6 h_T |v|_{2,\infty,T} \quad \forall v \in C^2(T). \tag{4.7}$$

For the proof see [1, pp. 85–87].

Remark 6.8. With a sliver tetrahedron (cf. Figure **(3)**)

$$A = (-h, 0, 0), \quad B = (0, h^3, -h), \quad C = (h, 0, 0), \quad D = (0, h^3, h),$$

we see that (4.6) holds, since $\varphi < \frac{\pi}{2}$, but (4.5) is violated for $h \to 0$. Similarly we observe that (4.6) is not valid and that (4.5) holds for $h \to 0$ if we consider a spike tetrahedron:

$$A = (0, 0, 0), \quad B = (h, 0, 0), \quad C = (h, 0, h^3), \quad D = (-h, h^3, 0).$$

These two examples show that conditions (4.5) and (4.6) are independent. In both the examples $\pi_T v$ loses its optimal order error behavior (4.7). See [36].

Remark 6.9. Theorem 6.6 shows that some badly-shaped tetrahedra preserve the optimal interpolation properties. They can therefore be safely used to fill narrow gaps and slots, see e.g. [1, p. 76], and also [37–43].

Remark 6.10. The maximum angle condition represents only a sufficient condition for the convergence of linear finite elements due to Theorem 6.6 and the famous Céa's lemma. According to [41], this condition is not necessary for the convergence of the FEM.

5   Discrete Maximum Principles for Linear Tetrahedral Finite Elements

The FEM uses piecewise polynomials to approximate solutions of partial differential equations. If these solutions satisfy certain maximum principles, it is desirable that their finite element approximations satisfy their discrete analogues (called discrete maximum principles, or DMPs in short). Nonobtuse and acute tetrahedral partitions indeed yield finite element approximations that satisfy DMPs for several elliptic [44–49] and parabolic problems [50–52] by means of continuous piecewise linear functions.

A key observation in this context is that the gradient of a non-zero linear function on a simplex T that vanishes on a face F_j of T is a constant non-zero normal to F_j. Hence, the sign of inner products between pairs of gradients of two distinct functions on T with this property is in one-to-one correspondence with the type of dihedral angle. To be more explicit, for $d \geq 1$ we have the following expression, which was derived in [44] directly from [5] and [49],

$$(\nabla v_i)^\top \nabla v_j = -\frac{\mathrm{vol}_{d-1} F_i \, \mathrm{vol}_{d-1} F_j}{(d \, \mathrm{vol}_d T)^2} \cos \alpha_{ij}, \quad i, j = 1, \dots, 4, \quad i \neq j, \tag{5.1}$$

where α_{ij} is the dihedral angle between F_i and F_j, and v_ℓ is the linear function that vanishes on F_ℓ and has value one at the vertex B_ℓ opposite F_ℓ (see Figure **(10)**). A similar expression was introduced in [53].

Basically, the discrete Laplacian that results from the standard finite element method has a non-negative inverse if each of the above inner products in the partition is non-positive for distinct i and j, which is the case for nonobtuse partitions. If the partition is in fact acute, the discrete Laplacian has a positive inverse and then reaction terms of small enough size can be handled using perturbation arguments. See for instance the papers [44, 54] where the presence of a reaction term in a reaction-diffusion problem led to the condition that the partition should be acute and the diameters of the simplices small enough.

Remark 6.11. It is also of practical interest that the DMP holds in order to avoid negative numerical values of nonnegative physical quantities like concentration, temperature, density, and pressure, see e.g. [45] for some real-life examples. Also a discrete heat flux may have an opposite sign than the continuous flux when the DMP is violated.

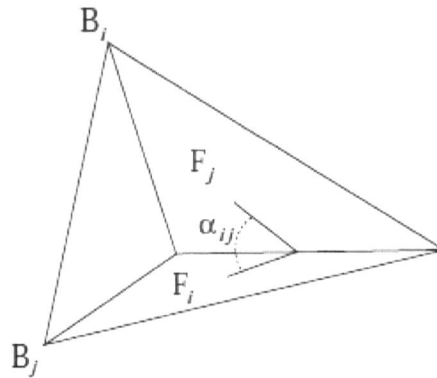

Figure **(10)**: Illustration for the above formula (5.1).

5.1   Nonobtuse tetrahedral partitions and their refinements

To increase the accuracy of FE calculations, we often need to perform various global or local refinements of the partitions. In this context, the techniques presented in Section 3 can be used. However, if we are interested in the preservation of the DMP on more refined partitions, then we should be able to guarantee the preservation of geometric properties of acuteness or nonobtuseness in the refining process.

For convenience, in Figure **(11)** we present several examples of nonobtuse tetrahedra, which are also mentioned in what follows. The left one, called a *path tetrahedron*, has three mutually orthogonal edges that form a path (in the sense of graph theory), the middle one, called a *cube corner tetraheron*, has three mutually orthogonal edges that share a common vertex.

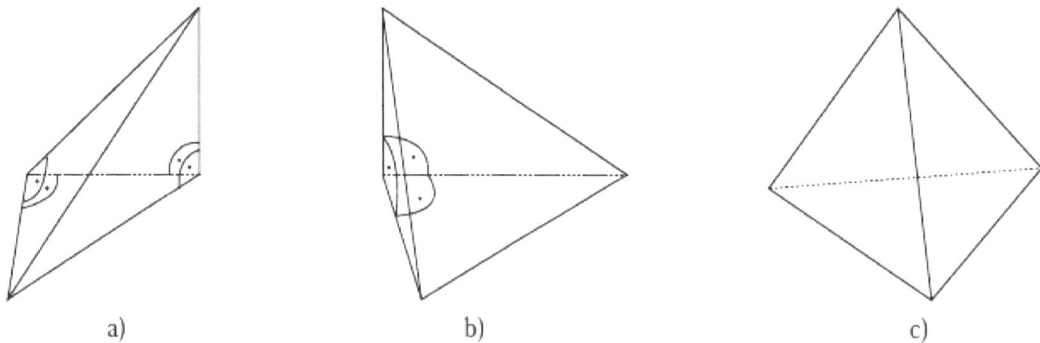

Figure **(11)**: Examples of nonobtuse tetrahedra: a) path, b) cube corner, and c) regular.

In [25], we presented sufficient conditions for the existence of partitions into path-tetrahedra with an arbitrarily small mesh size, as formulated in the following theorem.

Theorem 6.7. *Let each tetrahedron in a given nonobtuse partition of a polyhedron contain its circumcenter. Then there exists a family of partitions into path-tetrahedra.*

Its proof is constructive. Each face is first partitioned into four or six right triangles whose common vertex is the center of its circumscribed circle. Then each tetrahedron from the initial partition is divided into path-tetrahedra, by taking the convex hulls of the right triangles on its surface with its circumcenter (see Figure (**12**) (left)). Such a refinement technique is called *yellow* (cf. Figure (**6**)). In this case, common faces of adjacent tetrahedra from the initial partition are partitioned in the same manner. The proof then proceeds by induction.

Remark 6.12. In [55] the nonobtuseness assumption in Theorem 6.7 is replaced by a weaker condition that requires that only faces are nonobtuse. This enables us to apply the above technique also to degenerated tetrahedra (like needles, wedges, slivers, and splinters).

One technique for local nonobtuse tetrahedral refinements (towards a vertex) is presented in Figure (**12**), see [56] for details.

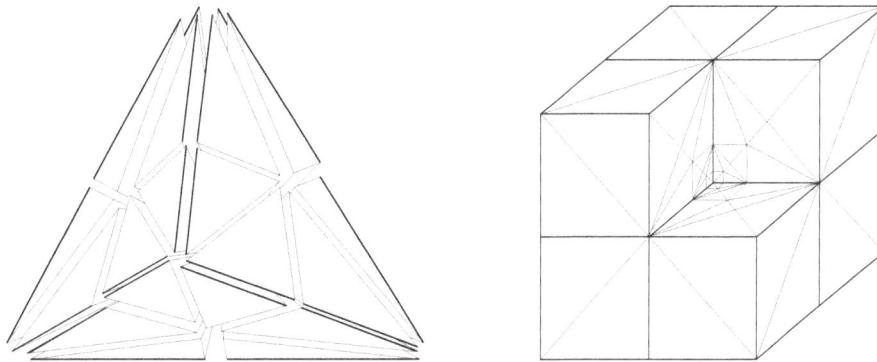

Figure (**12**): Global and local nonobtuse tetrahedral refinements from [25,56].

Further, we present the key idea and also an illustration from the recent work [57] (see Figures (**13**) and (**14**)) on nonobtuse tetrahedral refinements towards a flat face of (or interface inside) the solution domain. For this purpose we take a square prism (e.g. a cube) and its adjacent square prism. Denote their vertices and some other nodes as sketched in Figure (**13**), where also partitions of some faces are given.

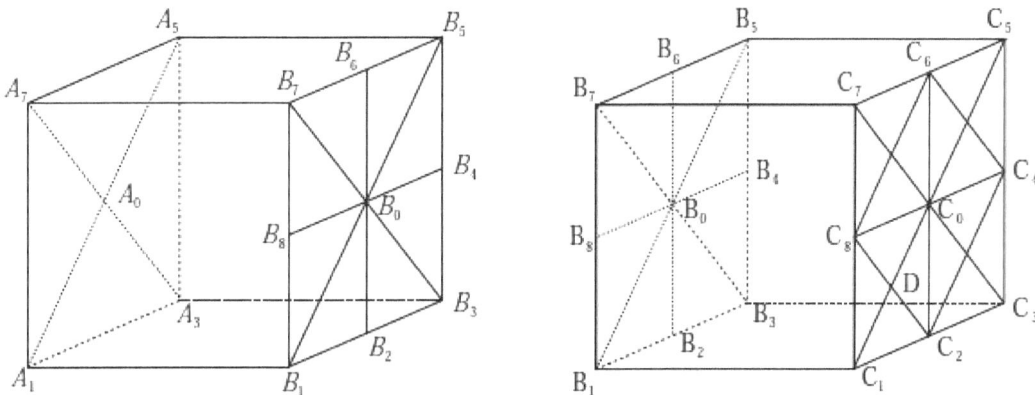

Figure (**13**): A sketch of a decomposition of two adjacent square prisms into nonobtuse tetrahedra.

In what follows, let $s = |B_1B_3| = |B_3B_5|$ denote the lengths of the edges of the square faces of the prisms, and let $l_1 = |A_0B_0|$ and $l_2 = |B_0C_0|$ be their thicknesses.

First, we decompose the left square prism $A_1A_3A_5A_7B_1B_3B_5B_7$ of Figure (**13**) into four triangular prisms whose common edge is A_0B_0. Second, we decompose each triangular prism into four tetrahedra. For instance, the triangular prism $A_0A_1A_3B_0B_1B_3$ will be divided in the following way (see Figure (**14**)):

$A_0A_1A_3B_0$ (cube corner tetrahedron), $A_1B_1B_2B_0$ (path tetrahedron),
$A_3B_3B_2B_0$ (path tetrahedron), and $A_1A_3B_0B_2$.

The first three resulting tetrahedra are clearly nonobtuse. The last tetrahedron $A_1A_3B_0B_2$ is the union of two path tetrahedra whose common face is $A_2B_0B_2$, where A_2 is the midpoint of A_1A_3. We see that $A_1A_3B_0B_2$ is nonobtuse if and only if

$$|B_1B_3| \leq 2|A_0B_0|, \quad \text{i.e.} \quad l_1 \geq \frac{s}{2}. \tag{5.2}$$

The other three triangular prisms, $A_0A_3A_5B_0B_3B_5$, $A_0A_5A_7B_0B_5B_7$, and $A_0A_1A_7B_0B_1B_7$, can be subdivided similarly.

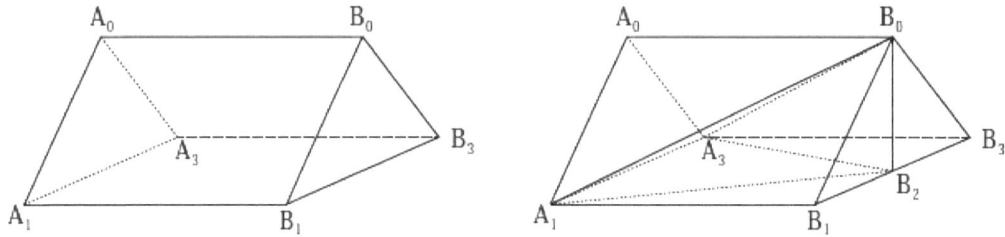

Figure (**14**): Decomposition of a triangular prism $A_0A_1A_3B_0B_1B_3$ into four tetrahedra.

Next, we decompose the right adjacent square prisms $B_1B_3B_5B_7C_1C_3C_5C_7$ of Figure (**13**) into eight triangular prisms whose common edge is B_0C_0. Further, the triangular prism $B_0B_1B_2C_0C_1C_2$ will be divided into four tetrahedra like in the previous step:

$B_0B_1B_2C_2$ (cube corner tetrahedron), $B_0C_0DC_2$ (path tetrahedron),
$B_1C_1DC_2$ (path tetrahedron), and $B_0B_1DC_2$.

The last tetrahedron is nonobtuse provided

$$|B_0B_1| \leq 2|B_0C_0|, \quad \text{i.e.} \quad l_2 \geq \frac{\sqrt{2}s}{4}. \tag{5.3}$$

This condition is necessary and sufficient to guarantee a nonobtuse decomposition of the triangular prism $B_0B_1B_2C_0C_1C_2$ into four nonobtuse tetrahedra as described above.

The other seven triangular prisms can be divided into nonobtuse tetrahedra similarly. In this way (i.e., under conditions (5.2) and (5.3)) we get a face-to-face nonobtuse partition of two adjacent square prisms. The left square prism of Figure (**13**) is subdivided into 16 and the right prism into 32 nonobtuse tetrahedra. This enables us to form layers and use this process repeatedly (see examples in [57]).

5.2 On acute tetrahedral partitions and their refinements

The following theorem states a relationship between dihedral angles and angles in triangular faces.

Theorem 6.8. *Let ABCD be an acute tetrahedron. Let α be the dihedral angle at the edge AD and let φ be the angle $\angle BAC$ with vertex at A. Then*

$$\varphi < \alpha. \tag{5.4}$$

For the proof, see [58, p. 384].

A similar theorem holds also for nonobtuse tetrahedra, but the inequality $<$ in (5.4) must be replaced by $\leq$. For obtuse tetrahedra, the inequality does not hold (cf. Remark 6.8). Such theorems can be used in the construction of acute and nonobtuse partitions of $\mathbf{R}^3$.

Remark 6.13. The first algorithm to partition the whole space $\mathbf{R}^3$ into acute tetrahedra was given in [59]. Later, in [60], four more algorithms were given, together with an acute tetrahedral partition of slabs. Recently, in [61], also the cube was partitioned into acute tetrahedra. Finally, in [62] all other Platonic solids were acutely partitioned.

Remark 6.14. Note that small enough perturbations of acute partitions remain acute. This is the not the case for nonobtuse partitions. Further properties of acute partitions are given in our survey paper [2].

6   Generalizations to Higher Dimensions

Several natural generalizations of the previous geometric results to higher dimensions are presented in below.

A *simplex* S in $\mathbf{R}^d$ is a convex hull of $d+1$ points, $A_1, A_2, \dots, A_{d+1}$, that do not belong to the same hyperplane. We denote by h_S the length of the longest edge of S. Let F_i be the $(d-1)$-dimensional facet of a simplex S opposite to the vertex A_i and let v_i be the altitude from the vertex A_i to the facet F_i. Formula (2.3) for the radius of the inscribed ball of S can be easily generalized to an arbitrary space dimension, namely

$$r_S = \frac{d \operatorname{vol}_d S}{\operatorname{vol}_{d-1} \partial S}. \tag{6.1}$$

By [63, 64], or [65, p. 125], the volume of a d-simplex S can be computed in terms of lengths of its edges using the so-called Cayley-Menger determinant of size $(d+2) \times (d+2)$

$$D_d = (-1)^{d+1} 2^d (d!)^2 (\operatorname{vol}_d S)^2 = \det \begin{bmatrix} 0 & 1 & 1 & \cdots & 1 & 1 \\ 1 & 0 & a_{12}^2 & \cdots & a_{1d}^2 & a_{1,d+1}^2 \\ 1 & a_{21}^2 & 0 & \cdots & a_{2d}^2 & a_{2,d+1}^2 \\ \vdots & \vdots & \vdots & \ddots & \vdots \\ 1 & a_{d+1,1}^2 & a_{d+1,2}^2 & \cdots & a_{d+1,d}^2 & 0 \end{bmatrix}, \tag{6.2}$$

where a_{ij} is the length of the edge $A_i A_j$ for $i \neq j$.

The radius R_S of the circumscribed ball B satisfies (see [66])

$$R_S^2 = -\frac{1}{2} \frac{\Delta_d}{D_d}, \tag{6.3}$$

where

$$\Delta_d = \det \begin{bmatrix} 0 & a_{12}^2 & \cdots & a_{1d}^2 & a_{1,d+1}^2 \\ a_{21}^2 & 0 & \cdots & a_{2d}^2 & a_{2,d+1}^2 \\ \vdots & \vdots & & \ddots & \vdots \\ a_{d+1,1}^2 & a_{d+1,2}^2 & \cdots & a_{d+1,d}^2 & 0 \end{bmatrix}.$$

Let $\Omega \subset \mathbf{R}^d$ be a domain. If the boundary of the closure $\partial\overline{\Omega}$ of $\overline{\Omega}$ is contained in a finite number of $(d-1)$-dimensional hyperplanes, we say that $\overline{\Omega}$ is *polytopic*. Moreover, if $\overline{\Omega}$ is bounded, it is called a *polytope*; in particular, $\overline{\Omega}$ is called a *polygon* for $d = 2$ and a *polyhedron* for $d = 3$.

We shall again consider only face-to-face simplicial partitions of a polytope $\overline{\Omega}$ and their families $\mathcal{F}$.

Condition 1′: There exists $c_1 > 0$ such that for any $\mathcal{T}_h \in \mathcal{F}$ and any $S \in \mathcal{T}_h$ we have

$$\text{vol}_d S \ge c_1 h_S^d. \tag{6.4}$$

Condition 2′: There exists $c_2 > 0$ such that for any $\mathcal{T}_h \in \mathcal{F}$ and any $S \in \mathcal{T}_h$ we have

$$\text{vol}_d b \ge c_2 h_S^d, \tag{6.5}$$

where $b \subset S$ is the inscribed ball of S.

Condition 3′: There exists $c_3 > 0$ such that for any $\mathcal{T}_h \in \mathcal{F}$ and any $S \in \mathcal{T}_h$ we have

$$\text{vol}_d S \ge c_3 \text{vol}_d B, \tag{6.6}$$

where $B \supset S$ is the circumscribed ball about S.

Condition 4′: There exists $c_4 > 0$ such that for any $\mathcal{T}_h \in \mathcal{F}$, any $S \in \mathcal{T}_h$, and any $i \in \{1, 2, \ldots, d+1\}$ we have

$$\sin_d(\hat{A}_i | A_1 A_2 \ldots A_{d+1}) \ge c_4, \tag{6.7}$$

where

$$\sin_d(\hat{A}_i | A_1 A_2 \ldots A_{d+1}) = \frac{d^{d-1} (\text{vol}_d S)^{d-1}}{(d-1)! \Pi_{j=1, j \ne i}^{d+1} \text{vol}_{d-1} F_j}. \tag{6.8}$$

Theorem 6.9. *Conditions 1′, 2′, 3′, and 4′ are equivalent.*

For the proof see [67] and [68]. If one of the conditions holds, then the family $\mathcal{F}$ of simplicial partitions is called *regular*.

Formula (5.1) can be rewritten as follows:

$$(\nabla v_i)^\top \nabla v_j = -\frac{\cos \alpha_{ij}}{h_i h_j}, \quad i, j = 1, \ldots, d+1, \quad i \ne j, \tag{6.9}$$

where h_i is the height in S above F_i and α_{ij} are dihedral angles between facets F_i and F_j. Their definition is similar to (2.6).

Many other results from the previous sections have been generalized to any dimension, for instance, local nonobtuse simplicial refinements towards a vertex [5], superconvergence phenomena [30], the maximum angle condition [42], the discrete maximum principle [44–47, 51].

Acknowledgments

The second author was supported by Grant MTM2008-03541 of the MICINN, Spain, the ERC Advanced Grant FP7-246775 NUMERIWAVES and Grant PI2010-04 of the Basque Government. The third author was supported by the Grant no. IAA 100190803 of the Grant Agency of the Academy of Sciences of the Czech Republic and the Institutional Research Plan AV0Z 10190503.

Bibliography

[1] Křížek M, Neittaanmäki P. Mathematical and numerical methods in modelling in electrical engineering: theory and applications. Kluwer Academic Publishers 1996.

[2] Brandts J, Korotov S, Křížek M, Šolc J. On nonobtuse simplicial partitions. SIAM Rev 2009; 51: 317-335.

[3] Rektorys K. Survey of applicable mathematics, Vol. II. Dordrecht: Kluwer 1994.

[4] Fiedler M. Geometrie simplexu v $\mathbf{E}_n$. Časopis Pěst Mat 1954; XII: 297-320.

[5] Brandts J, Korotov S, Křížek M. Dissection of the path-simplex in $\mathbf{R}^n$ into n path-subsimplices. Linear Algebra Appl 2007; 421: 382-393.

[6] Eriksson F. The law of sines for tetrahedra and n-simplices. Geom Dedicata 1978; 7: 71-80.

[7] Bern M, Chew P, Eppstein D, Ruppert J. Dihedral bounds for mesh generation in high dimensions. In: Proc. 6-th AMC-SIAM Sympos. on Discrete Algorithms 1995; pp. 189-196.

[8] Cheng SW, Dey TK, Edelsbrunner H, Facello MA, Teng SH. Sliver exudation. In: Proc. 15-th ACM Symp. Comp Geometry, 1999; pp. 1-13.

[9] Edelsbrunner H. Triangulations and meshes in computational geometry. Acta Numer 2000; 9: 133-213.

[10] Liu A, Joe B. Relationship between tetrahedron shape measures. BIT 1994; 34: 268-287.

[11] Parthasarathy VN, Graichen CM, Hathaway AF. A comparison of tetrahedron quality measures. Finite Elem Anal Des 1993; 15: 255-261.

[12] Golias NA, Tsiboukis TD. An approach to refining three-dimensional tetrahedral meshes based on Delaunay transformations. Internat J Numer Methods Engrg 1994; 37: 793-812.

[13] Schewchuk JR. What is a good linear finite element? Interpolation, conditioning, anisotropy, and quality measures, Preprint Univ. of California at Berkeley 2002; pp. 1-66.

[14] Křížek M. An equilibrium finite element method in three-dimensional elasticity. Apl Mat 1982; 27: 46-75.

[15] Bey J. Tetrahedral grid refinement. Computing 1995; 55: 355-378.

[16] Bey J. Simplicial grid refinement: on Freudenthal's algorithm and the optimal number of congruent classes. Numer Math 2000; 85: 1-29.

[17] Hannukainen A, Korotov S, Křížek M. On global and local mesh refinements by a generalized conforming bisection algorithm. J Comput Appl Math 2010; 235: 419-436.

[18] Zhang S. Successive subdivisions of tetrahedra and multigrid methods on tetrahedral meshes. Houston J Math 1995; 21: 541-556.

[19] Dauge M. Elliptic boundary value problems on corner domains. Smoothness and asymptotics of solutions. Lecture Notes in Mathematics, 1341; Berlin: Springer 1988.

[20] Grisvard P. Elliptic problems in nonsmooth domains. Monographs and Studies in Mathematics, 24, Pitman Harlow 1985.

[21] Bank RE. PLTMG: A software package for solving partial differential equations: users' guide 7.0. Philadelphia: SIAM 1994.

[22] Bänsch E. Local mesh refinement in 2 and 3 dimensions. IMPACT Comp Sci Engrg 1991; 3: 181-191.

[23] Felcman J, Dolejší V. Adaptive methods for the solution of the Euler equations in elements of blade machines. Z Angew Math Mech 1996; 76(Suppl 4): 301-304.

[24] Kornhuber R, Roitzsch R. On adaptive grid refinement in the presence of internal or boundary layers. Impact Comput Sci Engrg 1990; 2: 40-72.

[25] Korotov S, Křížek M. Acute type refinements of tetrahedral partitions of polyhedral domains. SIAM J Numer Anal 2001; 39: 724-733.

[26] Křížek M, Strouboulis T. How to generate local refinements of unstructured tetrahedral meshes satisfying a regularity ball condition. Numer Methods Partial Differential Equations 1997; 13: 201-214.

[27] Guo BQ. The *h-p* version of the finite element method for solving boundary value problems in polyhedral domains. In: Costabel M *et al.* , Eds. Proceedings of Boundary Value Problems and Integral Equations in Nonsmooth Domains (Luminy, 1993); Lecture Notes in Pure and Appl Math 167, Marcel Dekker, New York; 1995; pp. 101-120.

[28] Coxeter HSM. Trisecting an orthoscheme. Comput Math Appl 1989; 17: 59-71.

[29] Korotov S, Křížek M. Finite element analysis of variational crimes for a quasilinear elliptic problem in 3D. Numer Math 2000; 84: 549-576.

[30] Brandts J, Křížek M. Gradient superconvergence on uniform simplicial partitions of polytopes. IMA J Numer Anal 2003; 23: 489-505.

[31] Křížek M. Superconvergence phenomena on three-dimensional meshes. Int J Numer Anal Model 2005; 2: 43-56.

[32] Ciarlet PG. The finite element method for elliptic problems. Amsterdam: North-Holland 1978.

[33] Brandts J, Korotov S, Křížek M. On the equivalence of regularity criteria for triangular and tetrahedral finite element partitions. Comput Math Appl 2008; 55: 2227-2233.

[34] Zlámal M. On the finite element method. Numer Math 1968; 12: 394-409.

[35] Synge JL. The hypercircle in mathematical physics. Cambridge: Cambridge University Press 1957.

[36] Křížek M. On the maximum angle condition for linear tetrahedral elements. SIAM J Numer Anal 1992; 29: 513-520.

[37] Apel T. Anisotropic finite elements: local estimates and applications. Stuttgart: Advances in Numerical Mathematics, B. G. Teubner 1999.

[38] Babuška I, Aziz AK. On the angle condition in the finite element method. SIAM J Numer Anal 1976; 13: 214-226.

[39] Barnhill RE, Gregory JA. Sard kernel theorems on triangular domains with application to finite element error bounds. Numer Math 1975/1976; 25: 215-229.

[40] Gregory JA. Error bounds for linear interpolation on triangles. In: Whiteman JR, Ed. Proceedings of MAFELAP II. London: Academic Press 1976; pp. 163-170.

[41] Hannukainen A, Korotov S, Křížek M. The maximum angle condition is not necessary for the convergence of the finite element method (submitted in 2010).

[42] Jamet P. Estimations de l'erreur pour des éléments finis droits presque dégénérés. RAIRO Anal Numér 1976; 10: 43-60.

[43] Křížek M. On semiregular families of triangulations and linear interpolation. Appl Math 1991; 36: 223-232.

[44] Brandts J, Korotov S, Křížek M. The discrete maximum principle for linear simplicial finite element approximations of a reaction-diffusion problem. Linear Algebra Appl 2008; 429: 2344-2357.

[45] Karátson J, Korotov S. Discrete maximum principles for finite element solutions of nonlinear elliptic problems with mixed boundary conditions. Numer Math 2005; 99: 669-698.

[46] Karátson J, Korotov S. An algebraic discrete maximum principle in Hilbert space with applications to nonlinear cooperative elliptic systems. SIAM J Numer Anal 2009; 47: 2518-2549.

[47] Karátson J, Korotov S, Křížek M. On discrete maximum principles for nonlinear elliptic problems. Math Comput Simulation 2007; 76: 99-108.

[48] Korotov S, Křížek M, Neittaanmäki P. Weakened acute type condition for tetrahedral triangulations and the discrete maximum principle. Math Comp 2001; 70: 107-119.

[49] Křížek M, Lin Qun On diagonal dominance of stiffness matrices in 3D. East-West J Numer Math 1995; 3: 59-69.

[50] Faragó I, Horváth R. Discrete maximum principle and adequate discretizations of linear parabolic problems. SIAM J Sci Comput 2006; 28: 2313-2336.

[51] Faragó I, Karátson J, Korotov S. Discrete maximum principles for the FEM solution of some nonlinear parabolic problems. Electron Trans Numer Anal 2010; 36: 149-167.

[52] Fujii H. Some remarks on finite element analysis of time-dependent field problems. In: Theory and Practice in Finite Element Structural Analysis. Tokyo: University of Tokyo Press 1973; pp. 91-106.

[53] Xu J, Zikatanov L. A monotone finite element scheme for convection-diffusion equations. Math Comp 1999; 68: 1429-1446.

[54] Ciarlet PG, Raviart PA. Maximum principle and uniform convergence for the finite element method. Comput Methods Appl Mech Engrg 1973; 2: 17-31.

[55] Křížek M, Pradlová J. Nonobtuse tetrahedral partitions. Numer Methods Partial Differential Equations. 2000; 16: 327-334.

[56] Korotov S, Křížek M. Global and local refinement techniques yielding nonobtuse tetrahedral partitions. Comput Math Appl 2005; 50: 1105-1113.

[57] Korotov S, Křížek M. Nonobtuse local tetrahedral refinements towards a polygonal face/interface. Appl Math Letters (submitted in 2010), 1-6.

[58] Křížek M. There is no face-to-face partition of R^5 into acute simplices. Discrete Comput Geom 2006; 36: 381–390, Erratum 40 (2010).

[59] Üngör A. Tiling 3D Euclidean space with acute tetrahedra. In: Proceedings of Canadian Conference on Computational Geometry; 2001 Waterloo; pp. 169-172.

[60] Eppstein D, Sullivan JM, Üngör A. Tiling space and slabs with acute tetrahedra. Comput Geom: Theory and Appl 2004; 27: 237-255.

[61] VanderZee E, Hirani AN, Zharnitsky V, Guoy D. A dihedral acute triangulation of the cube. Comput Geom 2010; 43: 445-452.

[62] Kopczyński E, Pak I, Przytycki P. Acute triangulations of polyhedra and $\mathbf{R}^n$. arXiv:0909.3706 2009.

[63] Blumenthal LM. Theory and applications of distance geometry. Clarendon Press, Oxford, Chelsea, Publishing Co., New York, 1953, 1970.

[64] Ivanoff VF. The circumradius of a simplex. Math Magazine 1970; 43: 71-72.

[65] Sommerville DMY. An introduction to the geometry of n dimensions. New York: Dover Publications, Inc 1958.

[66] Berger M. Geometry, vol. 1, Berlin: Springer-Verlag 1987.

[67] Brandts J, Korotov S, Křížek M. On the equivalence of ball conditions for simplicial finite elements in $\mathbf{R}^d$. Appl Math Lett 2009; 22: 1210-1212.

[68] Brandts J, Korotov S, Křížek M. Generalization of the Zlámal condition for simplicial finite elements in $\mathbf{R}^d$. Appl Math 2011; 56 (to appear).

Chapter 7

Local Multigrid on Adaptively Refined Meshes and Multilevel Preconditioning with Applications

Ronald H.W. Hoppe[1], Xuejun Xu[2] and Huangxin Chen[3]

Abstract: We consider local multigrid methods for adaptive finite element and adaptive edge element discretized boundary value problems as well as multilevel preconditioned iterative solvers for the finite element discretization of a special class of saddle point problems. The local multigrid methods feature local smoothing processes on adaptively refined meshes and are applied to adaptive $P1$ conforming finite element discretizations of linear second order elliptic boundary value problems and to adaptive curl-conforming edge element approximations of H(curl)-elliptic problems and the time-harmonic Maxwell equations. On the other hand, the multilevel preconditioned iterative schemes feature block-diagonal or upper block-triangular preconditioned GMRES or BiCGStab applied to the resulting algebraic saddle point problems and preconditioned CG applied to the associated Schur complement system.

As technologically relevant applications of the above methods to problems in electromagnetism and acoustics, we consider the numerical simulation of Logging-While-Drilling tools in oil exploration and the numerical simulation of piezoelectrically actuated surface acoustic waves, respectively.

Keywords: *local multigrid methods, adaptively refined meshes, multilevel preconditioners, saddle point problems, Logging-While-Drilling, surface acoustic waves*

[1](corresponding author) Department of Mathematics, University of Houston, USA, and Institute of Mathematics, University at Augsburg, Germany; e-mail: rohop@math.uh.edu

[2]LSEC, Institute of Computational Mathematics, Chinese Academy of Sciences, Beijing, People's Republic of China; e-mail: xxj@lsec.cc.ac.cn

[3]LSEC, Institute of Computational Mathematics, Chinese Academy of Sciences, Beijing, People's Republic of China; e-mail: chx@lsec.cc.ac.cn

Owe Axelsson and János Karátson (Eds)

1 Introduction

Multigrid or multilevel and domain decomposition methods are the methods of choice when it comes to the efficient numerical solution of large linear systems arising from the finite element discretization of partial differential equations (cf., e.g., [1–7] and the references therein). For conforming finite elements on quasi-uniform meshes, the convergence properties of multigrid and multilevel methods have been further studied in [8–13]. A unified framework for a convergence analysis of multilevel and domain decomposition methods has been provided in [14] based on the notions of space decomposition and subspace correction.

On the other hand, during the past three decades adaptive finite element methods based on reliable and/or efficient a posteriori error estimators for local grid adaptation have been intensively studied and have reached some state of maturity as documented by a series of monographs (cf., e.g., [15–19]). For conforming adaptive finite element discretizations of linear second order elliptic boundary value problems, an overview on convergence results has been given in [20] and optimality has been addressed in [21–23]. The related issues for adaptive edge element discretizations of H(curl)-elliptic problems and the time-harmonic Maxwell equations have been studied in [24–27].

Since adaptive grid refinement techniques provide a hierarchy of meshes, it is natural to consider the application of multilevel techniques for adaptively generated meshes which actually has been initiated roughly twenty years ago. The approach in [28, 29] is the fast adaptive composite grid (FAC) method which uses global and local uniform grids both to define the composite grid problem and to interact for achieving a fast solution. Other approaches are the multilevel adaptive technique (MLAT) studied, e.g., in [30, 31] and multigrid methods for locally refined finite element meshes [32–36]. However, these locally refined meshes are subject to restrictive assumptions which are not met by the newest vertex bisection algorithm which is often used for refinement in the adaptive cycle consisting of the basic steps 'SOLVE', 'ESTIMATE', 'MARK', and 'REFINE'. The paper [37] was the first one to establish convergence of the multigrid V-cycle for nodal based finite element discretizations of linear second order elliptic problems without these restrictions and thus including the newest vertex bisection refinement strategy. The method features a local Gauss-Seidel smoother, i.e., a Gauss-Seidel iteration acting only on new nodes and those old nodes where the support of the associated nodal basis function has changed. Recently, optimality of such local multigrid methods has been shown in [38] based on the Schwarz theory well-known from the domain decomposition methodology [6].

This chapter is organized as follows. In section 2, we will be concerned with local multigrid methods for adaptive finite element discretizations of linear second order elliptic boundary value problems and adaptive edge element discretizations of boundary value problems for H(curl)-elliptic equations and the time-harmonic Maxwell equations. Level-independent multigrid convergence rates are derived within the Schwarz theory under assumptions that have to be verified for the local smoothers involved in the local multigrid methods. Section 3 is devoted to multilevel preconditioned iterative schemes for a special class of saddle point problems featuring block preconditioned GMRES and BiCGStab as well as preconditioned CG for the associated Schur complement system. The final sections 4 and 5 deal with technologically relevant applications. In particular, in section 4 we consider the numerical simulation of Logging-While-Drilling (LWD) tools that are used in oil exploration for measuring relevant geohydraulic parameters of the geological formation surrounding a borehole during the drilling process. For LWD tools with electromagnetic transmitters and receivers, the forward problem amounts to the solution of the time-harmonic Maxwell equations which can be solved using those local multigrid methods described in section 2. In section 5, we study the numerical solution of piezoelectrically actuated surface acoustic waves (SAW). SAW can be used, e.g., for signal processing in telecommunications or as nano-pumps in a microfluidic lab-on-a-chip. Such chips have their applications in clinical diagnostics, pharmacology, and forensics for high-throughput screening and hybridization in genomics, protein profiling in proteomics, and cytometry in cell analysis. The mathematical model gives rise to a saddle point problem of the form studied in section 3 and can thus be numerically solved by multilevel preconditioned iterative solvers.

2   Local Multigrid Methods on Adaptively Generated Meshes

In this section and throughout the rest of the chapter, we use standard notation from Lebesgue and Sobolev space theory. In particular, for a bounded domain $\Omega \subset \mathbb{R}^d, d \in \mathbb{N}$, we denote by $L^2(\Omega)$ and $\mathbf{L}^2(\Omega) := L^2(\Omega)^d$ the Hilbert spaces of square-integrable scalar- and vector-valued functions on Ω, respectively. Further, we denote by $H^1(\Omega)$ the Sobolev space of square integrable functions with square integrable weak derivatives equipped with the inner product $(\cdot, \cdot)_{1,\Omega}$ and norm $\| \cdot \|_{1,\Omega}$. For $\Sigma \subseteq \partial\Omega$, we refer to $H^{1/2}(\Sigma)$ as the space of traces $v|_{\Sigma}$ of functions $v \in H^1(\Omega)$ on Σ. We set $H^1_{0,\Sigma}(\Omega) := \{v \in H^1(\Omega) | v|_{\Sigma} = 0\}$ and refer to $H^{-1}_{\Sigma}(\Omega)$ as the associated dual space. For a simply connected polyhedral domain Ω with boundary $\Gamma = \partial\Omega$ we refer to $\mathbf{H}(\mathbf{curl};\Omega)$ as the Hilbert space of vector fields $\mathbf{q} \in \mathbf{L}^2(\Omega)$ such that $\nabla \times \mathbf{q} \in \mathbf{L}^2(\Omega)$, equipped with the standard graph norm $\| \cdot \|_{curl,\Omega}$. We denote by $\mathbf{H_0}(\mathbf{curl};\Omega)$ the subspace of vector fields with vanishing tangential trace components on Γ.

We assume V and H to be Hilbert spaces of functions on Ω with inner products $(\cdot, \cdot)_V, (\cdot, \cdot)_H$ and associated norms $\| \cdot \|_V, \| \cdot \|_H$ such that $V \subset H \subset V^*$ and V is continuously embedded in H. Given a bounded, V-elliptic bilinear form $a(\cdot, \cdot) : V \times V \to \mathbb{R}$ and a bounded linear functional $\ell \in V^*$, we consider the variational equation: Find $u \in V$ such that

$$a(u,v) = \ell(v) \quad , \quad v \in V. \tag{2.1}$$

In view of the Lemma of Lax-Milgram [39], the variational equation (2.1) admits a unique solution $u \in V$.

Example 1: A typical example is $V = H^1_0(\Omega), H = L^2(\Omega)$ and

$$a(u,v) = \int_{\Omega} \left(a\nabla u \cdot \nabla v + cuv \right) d\mathbf{x} \quad , \quad \ell(v) := \int_{\Omega} fv \, d\mathbf{x}, \tag{2.2}$$

where $f \in L^2(\Omega)$, $a = (a_{ij})^d_{i,j=1}, a_{ij} \in L^{\infty}(\Omega), 1 \leq i, j \leq d$, is a symmetric, uniformly positive definite matrix-valued function, and $c \in L^{\infty}_+(\Omega)$. Here, (2.1) represents the weak formulation of a second order elliptic boundary value problem.

Example 2: Another example is $V = \mathbf{H_0}(\mathbf{curl};\Omega), H = \mathbf{L}^2(\Omega)$ with

$$a(u,v) = \int_{\Omega} \left(a(\nabla \times u) \cdot (\nabla \times v) + cu \cdot v \right) d\mathbf{x} \quad , \quad \ell(v) := \int_{\Omega} f \cdot v \, d\mathbf{x}, \tag{2.3}$$

where $f \in \mathbf{L}^2(\Omega)$, $a \in L^{\infty}(\Omega)$ such that $a(x) \geq a_0 > 0$ a.e. in Ω, and $c \in L^{\infty}(\Omega)$. In case $c \in L^{\infty}_+(\Omega)$, the variational equation (2.1) is the weak formulation of an H(curl)-elliptic boundary value problem, e.g., arising from a semi-discretization in time of the eddy currents equations. On the other hand, if $c(x) < 0$ a.e. in Ω as it is the case for the Helmholtz problem associated with the time-harmonic Maxwell equations, the bilinear form $a(\cdot, \cdot)$ is not V-elliptic, but satisfies a Gårding-type inequality. Under suitable assumptions on the data it can be shown that for the solution of (2.1) a Fredholm alternative holds true (cf., e.g., [40]).

We assume $(V_i)_{i=0}^L, L \in \mathbb{N}, V_i = \text{span}\{\varphi_1^{(i)}, \cdots, \varphi_{N_i}^{(i)}\}, N_i \in \mathbb{N}, 0 \leq i \leq L$, to be a nested sequence $V_{i-1} \subset V_i, 1 \leq i \leq L$, of finite dimensional subspaces of V obtained, e.g., with respect to a nested hierarchy $\mathcal{T}_i(\Omega)$ of simplicial triangulations of Ω generated by the application of adaptive finite element methods to (2.1). For $D \subset \bar{\Omega}$ we refer to $\mathcal{N}_i(D), \mathcal{E}_i(D)$, and $\mathcal{F}_i(D)$ as the sets of nodes, edges, and faces of $\mathcal{T}_i(\Omega)$ in D. Moreover, for $E \in \mathcal{E}_i(D), F \in \mathcal{F}_i(D)$, and $T \in \mathcal{T}_i(\Omega)$ we denote by h_E the length of E, and by h_F and h_T the diameters of F and T, respectively. We set $h_i := \max\{h_T \mid T \in \mathcal{T}_i(\Omega)\}$.

The Galerkin approximation of (2.1) with respect to $V_i \subset V, 0 \leq i \leq L$, reads: Find $u_i \in V_i$ such that

$$a(u_i, v_i) = \ell(v_i) \quad , \quad v_i \in V_i. \tag{2.4}$$

If we define $A_i : V_i \to V_i$ by $(A_i u, v)_H = a(u, v), u, v \in V_i$, and $b_i \in V_i$ by $(b_i, v)_H = \ell(v), v \in V_i$, then 2.4 can be equivalently written as

$$A_i u_i = b_i. \tag{2.5}$$

Example 1: For (2.1) with $V = H_0^1(\Omega), H = L^2(\Omega)$ and the bilinear form $a(\cdot, \cdot)$ being given by (2.2), the natural choice for a finite element discretization is to choose nodal based conforming finite elements with respect to the simplicial triangulations $\mathcal{T}_i(\Omega)$ such as the Lagrangean finite elements of type (k) [39]. In particular, for $k = 1$ we obtain

$$V_i := \{v \in C_0(\Omega) \mid v|_T \in P_1(T) , T \in \mathcal{T}_h(\Omega)\}, \tag{2.6}$$

where $P_1(T)$ stands for the set of polynomials of degree 1 on T. The basis functions $\varphi_j^{(i)}, 1 \leq j \leq N_i$, are the nodal basis functions associated with the interior nodes $a_k \in \mathcal{N}_i(\Omega), 1 \leq k \leq N_i$, such that $\varphi_j^{(i)}(a_k) = \delta_{jk}, 1 \leq j, k \leq N_i$.

A hierarchy of adaptively refined meshes can be obtained, e.g., by residual-type a posteriori error estimators consisting of element residuals and edge residuals in 2D resp. element and face residuals in 3D (cf., e.g., [19]).

Example 2: In case $V = \mathbf{H_0}(\mathbf{curl}; \Omega), H = \mathbf{L}^2(\Omega)$ and $a(\cdot, \cdot)$ given by (2.3), a convenient choice for a curl-conforming finite element discretization are the edge elements of Nédélec's first family [41]

$$\mathbf{Nd^1}(T) := \{\mathbf{q} \mid \mathbf{q}(\mathbf{x}) = \mathbf{a} + \mathbf{b} \times \mathbf{x} , \mathbf{a}, \mathbf{b} \in \mathbb{R}^3\} , T \in \mathcal{T}_i(\Omega), \tag{2.7}$$

where each $\mathbf{q} \in \mathbf{Nd^1}(T)$ is uniquely determined by the zero moments of its tangential components on the six edges of T. This gives rise to the curl-conforming edge element spaces

$$V_i := \{v \in \mathbf{H_0}(\mathbf{curl}; \Omega) \mid v|_T \in \mathbf{Nd^1}(T) , T \in \mathcal{T}_i(\Omega)\}. \tag{2.8}$$

The basis functions $\varphi_j^{(i)}, 1 \leq j \leq N_i$, are the vector-valued functions associated with the interior edges $E_k \in \mathcal{E}_i(\Omega), 1 \leq k \leq N_i$, such that

$$h_{E_k}^{-1} \int_{E_k} \mathbf{t}_{E_k} \cdot \varphi_j^{(i)} \, ds = \delta_{jk} , 1 \leq j, k \leq N_i, \tag{2.9}$$

where $\mathbf{t}_{E_k}$ denotes the unit tangential vector on E_k.

Residual-type a posteriori error estimators for these edge element discretizations have been first studied in [42] and subsequently considered in [24–27].

We will solve (2.5) on level $i = L$ by local multigrid methods. As mentioned in the introductory section 1, local multigrid methods differ from standard multigrid schemes in so far as they feature local instead of global smoothing. To this end, we introduce

$$\mathcal{J}_i := \{1 \le j \le N_i \mid \nexists\, 1 \le j_{i-1} \le N_{i-1}\ \text{s.th.}\ \varphi_i^{(j)} = \varphi_{i-1}^{(j_{i-1})}\} \tag{2.10}$$

as the set of all indices $1 \le j \le N_i$ for which the level i basis function $\varphi_i^{(j)}$ does not correspond to a level $i-1$ basis function. Setting

$$\tilde{N}_i := \mathrm{card}(\mathcal{J}_i), \tag{2.11}$$

we rearrange the set of basis functions $\varphi_i^{(j)}, 1 \le j \le N_i$, according to

$$\{\varphi_i^{(1)}, \cdots, \varphi_i^{(\tilde{N}_i)}, \varphi_i^{(\tilde{N}_i+1)}, \cdots, \varphi_i^{(N_i)}\}, \tag{2.12}$$

such that $\varphi_i^{(j)}, 1 \le j \le \tilde{N}_i$, are the basis functions associated with the set $\mathcal{J}_i$ given by (2.10). For $1 \le i \le L$, we refer to $R_i : V_i \to V_i$ as a local smoothing operator that only operates on $\varphi_i^{(j)}, 1 \le j \le \tilde{N}_i$, whereas for $i = 0$ we choose $R_0 = A_0^{-1}$. We define projections $P_i, Q_i : V_L \to V_i, 0 \le i \le L-1$, by

$$a(P_i v, w) = a(v, w) \quad , \quad (Q_i v, w)_H = (v, w)_H \quad , \quad v \in V_L \, , \, w \in V_i. \tag{2.13}$$

Setting $V_i^{(j)} := \mathrm{span}\{\varphi_i^{(j)}\}, 1 \le j \le N_i$, we further define local projections $P_i^{(j)}, Q_i^{(j)} : V_L \to V_i^{(j)}$ and $A_i^{(j)} : V_i^{(j)} \to V_i^{(j)}$ according to

$$a(P_i^{(j)} v, \varphi_i^{(j)}) = a(v, \varphi_i^{(j)}) \quad , \quad (Q_i^{(j)} v, \varphi_i^{(j)})_H = (v, \varphi_i^{(j)})_H \quad , \quad v \in V_L, \tag{2.14}$$

$$(A_i^{(j)} v, \varphi_i^{(j)})_H = a(v, \varphi_i^{(j)}) \quad , \quad v \in V_i^{(j)}. \tag{2.15}$$

Then, the **local multigrid V-cycle** solves (2.5) by the iterative scheme

$$u_i^{(n+1)} = u_i^{(n)} + B_i(b_i - A_i u_i^{(n)}) \quad , \quad 0 \le i \le L \, , \, n \in \mathbb{N}_0. \tag{2.16}$$

Here, the operators $B_i, 0 \le i \le L$, are recursively given by $B_0 := A_0^{-1}$, whereas for $i \ge 1$ and $c \in V_i$ we define $B_i c = z_3$ with z_3 obtained by pre-smoothing, coarse-grid correction. and post-smoothing according to

Pre-smoothing: $z_1 = R_i b_i,$
Correction: $z_2 = z_1 + B_{i-1} Q_{i-1}(c - A_i z_1),$
Post-smoothing: $z_3 = z_2 + R_i(c - A_i z_2).$

Example 1: We consider the local Jacobi and the local Gauss-Seidel smoother. The local Jacobi smoother is an additive smoother given by

$$R_i := \gamma \sum_{j=1}^{\tilde{N}_i} (A_i^{(j)})^{-1} Q_i^{(j)}, \tag{2.17}$$

where $\gamma > 0$ is an appropriately chosen scaling parameter. On the other hand, the local Gauss-Seidel smoother is a multiplicative smoother given by

$$R_i := (I - E_i) A_i^{-1} \quad , \quad E_i := \prod_{j=1}^{\tilde{N}_i} (I - P_i^{(j)}). \tag{2.18}$$

Example 2: It is well known that for H(curl)-elliptic problems and the time-harmonic Maxwell equations the smoothing process has to take into account the non-trivial kernel of the discrete curl-operator which is given by the gradients of the nodal basis functions spanning the P_1-conforming finite element space (cf., e.g., [40]). In fact, one has to use a hybrid smoother which smoothes with respect to both the edge basis functions and the gradients of the nodal basis functions. Appropriate hybrid smoothers are the Hiptmair smoother [43, 44] and the Arnold-Falk-Winther smoother [45]. For local multigrid, the local version of the Hiptmair-Jacobi smoother is given as follows: We assume $N_i = \mathrm{card}(\mathcal{E}_i(\Omega)), M_i = \mathrm{card}(\mathcal{N}_i(\Omega))$ and refer to $\psi_i^{(j)}, 1 \leq j \leq N_i$, and $\theta_i^{(j)}, 1 \leq j \leq M_i$, as the edge and nodal basis functions, respectively. We define $\tilde{N}_i, \tilde{M}_i$ as in (2.10), (2.11) and set

$$\varphi_i^{(j)} := \left\{ \begin{array}{ll} \psi_i^{(j)}, & 1 \leq j \leq \tilde{N}_i \\ \nabla \theta_i^{(j-\tilde{N}_i)}, & \tilde{N}_i + 1 \leq j \leq \tilde{N}_i + \tilde{M}_i \end{array} \right. . \tag{2.19}$$

Then, the local Hiptmair-Jacobi smoother is the additive smoother given by

$$R_i := \gamma \sum_{j=1}^{\tilde{N}_i + \tilde{M}_i} (A_i^{(j)})^{-1} Q_i^{(j)}, \tag{2.20}$$

where $\gamma > 0$ is a scaling factor. The multiplicative Hiptmair-Gauss-Seidel smoother can be defined analogously.

The optimality of the local multigrid method in terms of level-independent convergence rates can be shown based on the well-known Schwarz theory as described, e.g., in [6, 13, 14]. For this purpose, we define operators $T : V_L \to V_L$ and $T_i : V_L \to V_i, 0 \leq i \leq L$, according to

$$T := \sum_{i=0}^{L} T_i \quad , \quad T_i := R_i A_i P_i , \ 0 \leq i \leq L. \tag{2.21}$$

The convergence of the local multigrid method will be measured in terms of the error operator

$$E := \prod_{i=0}^{L} (I - T_i), \tag{2.22}$$

where I stands for the identity in V_L.

Theorem 7.1. *We suppose that the operators $T_i, 0 \leq i \leq L$, and T satisfy the following assumptions:*

$(\mathbf{A_1})$: *The operators $T_i, 0 \leq i \leq L$, are nonnegative with respect to the inner product $a(\cdot, \cdot)$, and there exist constants $0 < \omega_i < 2, 0 \leq i \leq L$, such that for all $v \in V_L$*

$$a(T_i v, T_i v) \ \leq \ \omega_i \, a(T_i v, v) \quad , \quad 0 \leq i \leq L. \tag{2.23}$$

$(\mathbf{A_2})$: *There exists a stability constant $C_0 > 0$ such that for all $v \in V_L$*

$$a(v, v) \ \leq \ C_0 \, a(Tv, v). \tag{2.24}$$

$(\mathbf{A_3})$: *There exist constants $C_1, C_2 > 0$ such that for all $v, w \in V_L$*

$$\sum_{i=0}^{L} \sum_{j=0}^{i-1} a(T_i v, T_j w) \ \leq \ C_1 \left(\sum_{i=0}^{L} a(T_i v, v) \right)^{1/2} \left(\sum_{i=0}^{L} a(T_i w, w) \right)^{1/2}, \tag{2.25a}$$

$$\sum_{i=0}^{L} a(T_i v, w) \ \leq \ C_2 \left(\sum_{i=0}^{L} a(T_i v, v) \right)^{1/2} \left(\sum_{i=0}^{L} a(T_i w, w) \right)^{1/2}. \tag{2.25b}$$

Under these assumptions, the local multigrid method converges with

$$a(Ev, Ev) \leq \gamma \, a(v, v) \quad , \quad v \in V_L, \tag{2.26}$$

where $\gamma := 1 - (2 - \omega)/(\max(2, C_0(C_1 + C_2)^2))$, $\omega := \max_{0 \leq i \leq L} \omega_i$.

Proof. We refer to [6, 14] or [13]. ∎

In order to apply Theorem 7.1 to the local multigrid methods from Example 1 and Example 2 above, one has to verify the assumptions $(\mathbf{A_1})$,$(\mathbf{A_2})$, and $(\mathbf{A_3})$ for the respective local smoothers. As far as the local Jacobi and local Gauss-Seidel smoothers from Example 1 are concerned, this has been done in [38]. For the local Hiptmair-Jacobi and local Hiptmair-Gauss-Seidel smoothers from Example 2, similar arguments can be applied.

3   Multilevel Preconditioning of Saddle Point Problems

We assume V, H, and W to be Hilbert spaces of real- or complex-valued functions with inner products $(\cdot, \cdot)_V, (\cdot, \cdot)_H, (\cdot, \cdot)_W$ and associated norms $\| \cdot \|_V, \| \cdot \|_Q, \| \cdot \|_W$ such that $V \subset H \subset V^*$ and V is compactly embedded in H. We further suppose that $a(\cdot, \cdot) : V \times V \to \mathbb{K}$, $\mathbb{K} = \mathbb{R}$ or $\mathbb{K} = \mathbb{C}$, is a bounded, symmetric (resp. Hermitean) and V-elliptic bilinear (resp. sesquilinear) form, $b(\cdot, \cdot) : W \times V \to \mathbb{K}$ is a bounded bilinear (resp. sesquilinear) form, and $c(\cdot, \cdot) : W \times W \to \mathbb{K}$ is a bounded, symmetric (resp. Hermitean) and W-elliptic bilinear (resp. sesquilinear) form. We set $a_\omega(\cdot, \cdot) := a(\cdot, \cdot) - \omega^2(\cdot, \cdot)_H$, where $\omega \in \mathbb{R}_+$. Given bounded linear functionals $\ell_1 : V \to \mathbb{K}$ and $\ell_2 : W \to \mathbb{K}$, we consider saddle point problems of the form: Find $(u, w) \in V \times W$ such that

$$a_\omega(u, v) + b(w, v) = \ell_1(v) \quad , \quad v \in V, \tag{3.1a}$$
$$b(u, z) - c(w, z) = \ell_2(z) \quad , \quad z \in W. \tag{3.1b}$$

In case $\omega = 0$ and $\mathbb{K} = \mathbb{R}$, such problems arise, e.g., from the mixed formulation of elliptic boundary value problems and the weak formulation of the Stokes problem, where typically $c(\cdot, \cdot) = 0$ (cf., e.g., [46]), whereas for $\omega > 0$ they occur within the context of time-harmonic acoustics or time-harmonic electromagnetism ($\mathbb{K} = \mathbb{C}$) (cf., e.g., [40]). In section 5, we will deal with a problem representing the weak formulation of a model for piezoelectrically actuated surface acoustic waves.

We denote by $A : V \to V^*, B : W \to V^*$, and $C : W \to W^*$ the operators associated with the bilinear (sesquilinear) forms and by I the injection $I : V \to V^*$. Then, an equivalent formulation of (3.1a),(3.1b) is

$$(A - \omega^2 I)u + Bw = \ell_1 , \tag{3.2a}$$
$$B^* u - Cw = \ell_2 , \tag{3.2b}$$

where $B^* : V \to W^*$ stands for the adjoint of B. In particular, the operator A is self-adjoint and V-elliptic, and the operator C is self-adjoint and W-elliptic. In the sequel, we focus on the case where the operator C is invertible. Then, an elimination of w from (3.2a),(3.2b) results in the Schur complement system

$$(S - \omega^2 I)u = \ell. \tag{3.3}$$

Here, the operator $S : V \to V^*$ is defined according to

$$S := A + BC^{-1}B^*, \tag{3.4}$$

whereas the right-hand side ℓ is given by

$$\ell := \ell_1 + BC^{-1}\ell_2. \tag{3.5}$$

Theorem 7.2. *If $S^{-1} : Q \to V$ is a Hilbert-Schmidt operator, there holds:*

(i) *The spectrum of S consists of a sequence of countably many real eigenvalues $0 < \zeta_1^2 < \zeta_2^2 < \dots$ tending to infinity, i.e., $\lim_{j \to \infty} \zeta_j^2 = \infty$.*

(ii) *If ω^2 is not an eigenvalue of S, for every $\ell \in V^*$, (3.3) admits a unique solution $u \in V$ depending continuously on ℓ.*

(iii) *If ω^2 is an eigenvalue of S, (3.3) is solvable if and only if $\ell \in Ker(S - \omega^2 I)^0$ where*

$$Ker(S - \omega^2 I)^0 := \{ v^* \in V^* \mid \langle v^*, v \rangle = 0 \,,\, v \in Ker(S - \omega^2 I) \}.$$

Proof. The assertions $(i), (ii)$ and (iii) follow from the Hilbert-Schmidt theory and the Fredholm alternative (cf., e.g., [47]). ∎

Corollary 7.1. *If ω^2 is not an eigenvalue of S, then the operator $S_\omega := S - \omega^2 I$ satisfies the inf-sup condition*

$$\inf_{0 \neq u \in V} \sup_{0 \neq v \in V} \frac{|\langle S_\omega u, v \rangle|}{\|u\|_V \|v\|_V} \geq \beta > 0. \tag{3.6}$$

Proof. We refer to [46]. ∎

Remark 7.1. The constant β in the inf-sup condition (3.6) deteriorates when ω^2 comes close to an eigenvalue of S, which would have a negative impact on the convergence behavior of iterative schemes for the numerical solution of the discrete Schur complement system. However, for the application considered in section 5, ω^2 is sufficiently far away from an eigenvalue of S so that the convergence rates will not be affected by the actual value of β.

Given a null sequence $\mathcal{H}$ of positive real numbers, we assume $(V_h)_{h \in \mathcal{H}}, V_h \subset V, h \in \mathcal{H}$, and $(W_h)_{h \in}$, $W_h \subset W, h \in \mathcal{H}$, to be sequences of finite dimensional subspaces that are limit dense in V and W, respectively. The Galerkin approximation of the saddle point problem (3.1a),(3.1b) amounts to the computation of $(u_h, w_h) \in V_h \times W_h$ such that

$$a_\omega(u_h, v_h) + b(w_h, v_h) = \ell_1(v_h) \quad, \quad v_h \in V_h, \tag{3.7a}$$
$$b(z_h, u_h) - c(w_h, z_h) = \ell_2(z_h) \quad, \quad w_h \in W_h. \tag{3.7b}$$

We denote by $A_h : V_h \to V_h^*, B_h : W_h \to V_h^*, C_h : W_h \to W_h^*$ the operators associated with the restrictions $a|_{V_h \times V_h}, b|_{W_h \times V_h}, c|_{W_h \times W_h}$, and by I_h the injection $I_h : V_h \to V_h^*$, and we further define $\ell_{1,h} \in V_h^*$ and $\ell_{2,h} \in W_h^*$ analogously. Then, the operator form of (3.7a),(3.7b) reads as follows:

$$(A_h - \omega^2 I_h)u_h + B_h w_h = \ell_{1,h} \,, \tag{3.8a}$$
$$B_h^* u_h - C_h w_h = \ell_{2,h} \,. \tag{3.8b}$$

Static condensation of w_h gives rise to the discrete Schur complement system

$$(S_h - \omega^2 I_h)u_h = \ell_h , \tag{3.9}$$

where S_h and the right-hand side ℓ_h are given by

$$S_h := A_h + B_h C_h^{-1} B_h^* \quad , \quad \ell_h := \ell_{1,h} + B_h C_h^{-1} \ell_{2,h}.$$

Theorem 7.3. *Assume that ω^2 is not an eigenvalue of S as given by (3.4). Then, for sufficiently small h the discrete Schur complement system (3.9) admits a unique solution $u_h \in V_h$.*

Proof. If ω^2 is not an eigenvalue of S, the inf-sup condition (3.6) holds true which implies

$$\beta \|u\|_V \leq \sup_{v \in V \setminus \{0\}} \frac{|\langle (S - \omega^2 I)u, v \rangle|}{\|v\|_V} = \tag{3.10}$$

$$= \sup_{v \in V \setminus \{0\}} \frac{|\langle (S(u - \omega^2 S^{-1}u)), v \rangle|}{\|v\|_V} \leq \|S\| \|u - \omega^2 S^{-1}u\|_V.$$

On the other hand, we note that S_h is the Galerkin approximation of S, i.e.,

$$\langle S_h u_h, v_h \rangle = \langle S u_h, v_h \rangle \quad , \quad u_h, v_h \in V_h .$$

Hence, referring to $\alpha_S > 0$ as the ellipticity constant of S, we have

$$\langle S_h v_h, v_h \rangle \geq \alpha_S \|v_h\|_V^2 \quad , \quad v_h \in V_h. \tag{3.11}$$

Using (3.11), we deduce from (3.10) that

$$\sup_{0 \neq v_h \in V_h} \frac{|\langle (S_h - \omega^2 I_h)u_h, v_h \rangle|}{\|v_h\|_V} = \sup_{0 \neq v_h \in V_h} \frac{|\langle S_h(u_h - \omega^2 S_h^{-1}u_h), v_h \rangle|}{\|v_h\|_V}$$

$$\geq \frac{|\langle S_h(u_h - \omega^2 S_h^{-1}u_h), u_h - \omega^2 S_h^{-1}u_h \rangle|}{\|u_h - \omega^2 S_h^{-1}u_h\|_V} \geq \alpha_S \|u_h - \omega^2 S_h^{-1}u_h\|_V$$

$$\geq \alpha_S \left(\|u_h - \omega^2 S^{-1}u_h\|_V - \omega^2 \|(S_h^{-1} - S^{-1})u_h\|_V \right) \geq \beta_h \|u_h\|_V,$$

where

$$\beta_h := \alpha_S \left(\frac{\beta}{\|S\|} - \omega^2 \|S_h^{-1} - S^{-1}\| \right).$$

Due to the fact that S_h is elliptic uniformly in h and the Galerkin approximation of S, we have $S_h^{-1} \to S^{-1}$ as $h \to 0$. Hence, there exists $h_{max} > 0$ such that $\beta_h \geq \gamma > 0$ uniformly for $h \leq h_{max}$. This shows that $S_h - \omega^2 I_h$ satisfies a discrete inf-sup condition asymptotically as $h \to 0$ which gives the assertion. ∎

The discrete saddle point problem (3.7a),(3.7b) can be written equivalently as the algebraic saddle point problem

$$\begin{pmatrix} A & B \\ B^* & -C \end{pmatrix} \begin{pmatrix} u \\ w \end{pmatrix} = \begin{pmatrix} b_1 \\ b_2 \end{pmatrix}, \tag{3.12}$$

where $A \in \mathbb{R}^{n_h \times n_h}, n_h := \dim \mathbf{V}_h$, and $C \in \mathbb{R}^{m_h \times m_h}, m_h := \dim W_h$, are symmetric positive definite matrices, $B \in \mathbb{R}^{n_h \times m_h}$, and $b_1 \in \mathbb{R}^{n_h}, b_2 \in \mathbb{R}^{m_h}$. The algebraic saddle point problem (3.12) can be solved by preconditioned GMRES or BiCGStab [48, 49] using an upper block-triangular preconditioner P of the form

$$P = \begin{pmatrix} \tilde{A} & \tilde{B} \\ 0 & -\tilde{C} \end{pmatrix}$$

such that

$$\gamma_A v^T \tilde{A} v \leq v^T A v \leq \Gamma_A v^T \tilde{A} v, \qquad \gamma_C w^T \tilde{C} w \leq w^T C w \leq \Gamma_C w^T \tilde{C} w,$$

with constants $0 < \gamma_A \leq \Gamma_A, 0 < \gamma_C \leq \Gamma_C$ satisfying $\Gamma_A / \gamma_A \ll \kappa(A), \Gamma_C / \gamma_C \ll \kappa(C)$, where $\kappa(A), \kappa(C)$ are the spectral radii of A and C, respectively (cf., e.g., [50]). Alternatively, preconditioned CG [51] can be applied to the Schur complement system associated with (3.12).

In practical applications, where V, W are spaces of functions on a spatial domain $\Omega \subset \mathbb{R}^d$ and $V_i, W_i, 0 \leq i \leq L$, are finite element spaces with respect to a hierarchy $\{\mathcal{T}_i\}_{i=0}^L$ of triangulations of Ω, the operators $\tilde{A}^{-1}$ and $\tilde{C}^{-1}$, needed for the implementation of the preconditioned iterative scheme, can be realized, e.g., by BPX preconditioners [52]. Corresponding results within the context of the numerical simulation of piezoelectrically actuated surface acoustic waves will be reported in the subsequent section 5.

4   Numerical Simulation of LWD (Logging-While-Drilling) Tools

LWD (Logging-While-Drilling), sometimes also referred to as MWD (Measurements-While-Drilling), are techniques for the measurement of geological formation parameters such as resistivity and porosity during the excavation of boreholes, e.g., in deepwater drilling (cf. Figure (1) (left)). LWD uses tools that are integrated into the BHA (Bottom-Hole Assembly). The BHA is the lower part of the drillstring which consists of the bit, a mud motor for directional drilling, stabilizers, the drill collar, and the drillpipe (cf. Figure (1) (right)).

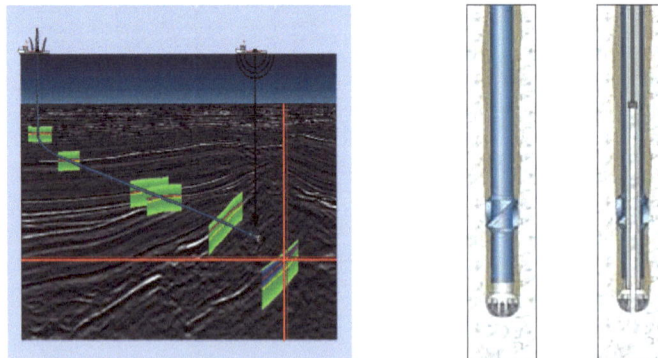

Figure (1): Horizontal deepwater drilling (l.) and a typical bottom-hole assembly (r.)

LWD tools based on an electromagnetic induction sensor are featuring saddle type transmitter and receiver antennas that are placed concentrically on a metallic mandrel (cf. Figure (2) (left)). The sensor is placed in the borehole with its axis being parallel to the borehole. The mandrel is a circular cylinder which is considered as a perfect electric conductor. The transmitter and receiver antennas with an aperture of 90° are imbedded in a sleeve and protected by a magnetic shielding (cf. Figure (2) (right)).

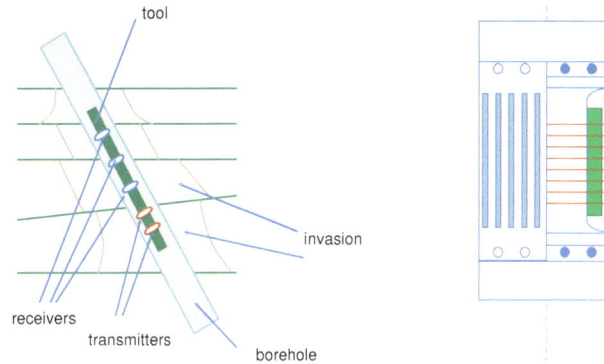

Figure **(2)**: Schematic representations of an LWD tool with two transmitter and three receiver antennas (l.) and a typical antenna configuration (r.). Courtesy of [53].

The transmitter antennas carry a current of $1A$. The frequency dependence is $\exp(-2\pi ift)$ with a frequency f up to 2 MHz. Typical dimensions of a saddle type antenna are shown in Figure **(3)** (left)). The problem to compute the open-circuit voltages features high conductivity contrasts (cf. Figure **(3)** (right)). The conductivity is 10^7 S/m in the mandrel, varies from 10 S/m to 10^{-3} S/m in the mud between the mandrel and the wall of the borehole, and ranges from 10^{-4} S/m to 10 S/m in the formation surrounding the borehole [53].

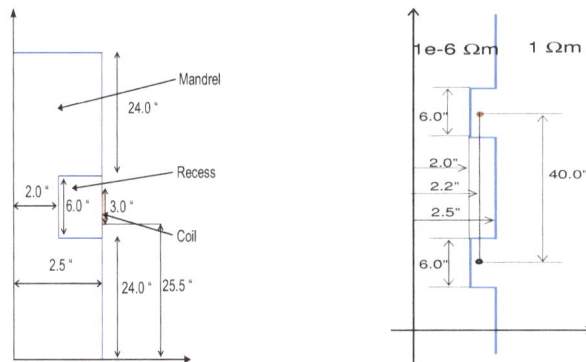

Figure **(3)**: Schematic representations of an LWD tool with two transmitter and three receiver antennas (l.) and a typical antenna configuration (r.). Courtesy of [53].

The computation of the geological formation parameters based on the data obtained at the receiver antennas amounts to the solution of an inverse scattering problem. Here, for an induction sensor with one transmitter and two receiver antennas we only consider part of the forward problem, namely the computation of the electric field $\mathbf{E}$ in a cylindrical domain $\Omega \subset \mathbb{R}^3$ between the mandrel and the wall of the borehole (cf. Figure **(4)**).

The boundary is split according to $\Gamma = \overline{\Gamma}_1 \cup \overline{\Gamma}_2 \cup \overline{\Gamma}_3, \Gamma_1 \cap \Gamma_2 \cap \Gamma_3 = \emptyset$, where Γ_1 represents the position of the transmitter antennas, Γ_2 stands for the wall of the borehole, and $\Gamma_3 := \Gamma \setminus (\overline{\Gamma}_1 \cup \overline{\Gamma}_2)$. Assuming a time-periodic excitation, the computation of the electric field $\mathbf{E}$ requires the solution of the time-

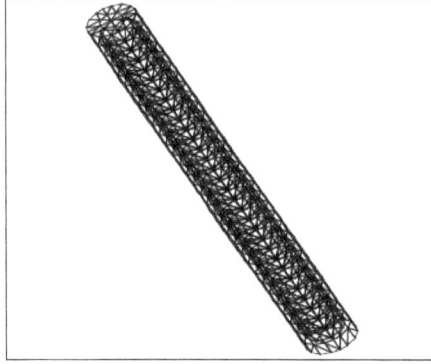

Figure **(4)**: Computational domain consisting of the cylindrical region between the mandrel and the walls of the borehole and its initial simplicial triangulation

harmonic Maxwell equations:

$$\nabla\times(\mu_r^{-1}\nabla\times\mathbf{E}) - \kappa^2\varepsilon_r\mathbf{E} = \mathbf{0} \quad \text{in } \Omega, \tag{4.1a}$$

$$(\nu\times\mathbf{E})\times\nu = \mathbf{g}_1 \quad \text{on } \Gamma_1, \tag{4.1b}$$

$$\nu\times(\mu_r^{-1}\nabla\times\mathbf{E}) - i\kappa\lambda\,\nu\times\mathbf{E} = \mathbf{g}_2 \quad \text{on } \Gamma_2, \tag{4.1c}$$

$$\nu\times(\mu_r^{-1}\nabla\times\mathbf{E}) = \mathbf{0} \quad \text{on } \Gamma_3. \tag{4.1d}$$

Here, ε_r and μ_r stand for the relative permittivity and relative permeability

$$\varepsilon_r = \frac{1}{\varepsilon_0}\left(\varepsilon + \frac{i\sigma}{\omega}\right) \quad , \quad \mu_r = \frac{\mu}{\mu_0}, \tag{4.2}$$

where ε, μ denote the permittivity and permeability of the medium and ε_0, μ_0 are the permittivity and permeability in vacuum. Moreover, σ refers to the conductivity and κ stands for the wavenumber $\kappa = \omega\sqrt{\varepsilon_0\mu_0}$, where $\omega = 2\pi f$ is the angular frequency. Finally, λ is given by

$$\lambda = (1+i)\sqrt{\frac{\pi\sigma}{\omega\mu}}, \tag{4.3}$$

and ν stands for the unit exterior normal vector on Γ. The tangential vector fields $\mathbf{g}_1$ and $\mathbf{g}_2$ are assumed to be given on Γ_1 and Γ_2, respectively. We refer to [40] for the derivation of (4.1a)-(4.1d) from Maxwell's equations.

As in section 2, we denote by $\mathbf{H}(\mathbf{curl};\Omega)$ the Hilbert space of complex-valued vector fields $\mathbf{q}$ with components in $L^2(\Omega)$ such that the components of $\nabla\times\mathbf{q}$ also live in $L^2(\Omega)$, equipped with the standard graph norm $\|\cdot\|_{curl,\Omega}$. We recall that the space $\mathbf{H}^{-1/2}(\mathbf{curl}_{\Gamma_i};\Gamma_i)$ is the trace space of tangential component traces $(\nu\times\mathbf{q})\times\nu$ on Γ_i, whereas the space $\mathbf{H}^{-1/2}(\mathbf{curl}_{\Gamma_i};\Gamma_i)$ stands for the trace space of tangential traces $\nu\times\mathbf{q}$ on Γ_2. Here, $\mathbf{curl}_{\Gamma_i}$ and div_{Γ_i} are the surfacic divergence and surfacic rotation, respectively (cf., e.g., [54]). Assuming

$$\mathbf{g}_1 \in \mathbf{H}^{-1/2}(\mathbf{curl}_{\Gamma_1};\Gamma_1) \quad , \quad \mathbf{g}_2 \in \mathbf{H}^{-1/2}(\text{div}_{\Gamma_2};\Gamma_2), \tag{4.4}$$

we set

$$\mathbf{V} := \{\mathbf{q} \in \mathbf{H}(\mathbf{curl};\Omega) \mid \mathbf{q}\times\nu|_{\Gamma_1} = \mathbf{g}_1\} \quad , \quad \mathbf{V}_0 := \{\mathbf{q} \in \mathbf{H}(\mathbf{curl};\Omega) \mid \mathbf{q}\times\nu|_{\Gamma_1} = \mathbf{0}\}.$$

The weak formulation of (4.1a)-(4.1d) is to find $\mathbf{E} \in \mathbf{V}$ such that

$$a_\Omega(\mathbf{E}, \mathbf{q}) + b_{\Gamma_2}(\mathbf{E}, \mathbf{q}) = \ell(\mathbf{q}) \quad , \quad \mathbf{q} \in \mathbf{V_0}. \tag{4.5}$$

Here, the sesquilinear forms $a_\Omega(\cdot, \cdot)$, $b_{\Gamma_2}(\cdot, \cdot)$, and the functional $\ell(\cdot)$ are given by

$$a_\Omega(\mathbf{E}, \mathbf{q}) := \int_\Omega \left(\mu_r^{-1}(\nabla \times \mathbf{E}) \cdot (\nabla \times \mathbf{q}) - (\kappa^2 \varepsilon_r + i\omega\sigma)\mathbf{E} \cdot \mathbf{q} \right) d\mathbf{x},$$

$$b_{\Gamma_2}(\mathbf{E}, \mathbf{q}) := \langle i\kappa\lambda\nu \times \mathbf{E}, (\nu \times \mathbf{q}) \times \nu \rangle_{\Gamma_2} \quad , \quad \ell(\mathbf{q}) := \langle \mathbf{g}_2, (\nu \times \mathbf{q}) \times \nu \rangle_{\Gamma_2},$$

where $\langle \cdot, \cdot \rangle_{\Gamma_2}$ is the dual pairing between $\mathbf{H}^{-1/2}(\mathrm{div}_{\Gamma_2}; \Gamma_2)$ and $\mathbf{H}^{-1/2}(\mathbf{curl}_{\Gamma_2}; \Gamma_2)$.

For sufficiently regular data of the problem, it is well-known that if κ is not an eigenvalue of the associated Maxwell eigenproblem, then the variational equation (4.5) has a unique solution $\mathbf{E} \in \mathbf{V}$ (cf., e.g., [40]).

Given a simplicial triangulation $\mathcal{T}_h(\Omega)$ of the computational domain Ω that aligns with the partition of the boundary Γ, we discretize (4.5) by the lowest order edge elements

$$\mathbf{Nd}^1(T) := \{\mathbf{q} \mid \mathbf{q}(\mathbf{x}) = \mathbf{a} + \mathbf{b} \times \mathbf{x}, \ \mathbf{a}, \mathbf{b} \in \mathbb{R}^3\},$$

of Nédélec's first family [41] with the degrees of freedom given by zero order moments of the tangential trace components on the six edges of $T \in \mathcal{T}_h(\Omega)$. We refer to

$$\mathbf{Nd}^1(\Omega; \mathcal{T}_h(\Omega)) := \{\mathbf{q}_h \in \mathbf{H}(\mathbf{curl}; \Omega) \mid \mathbf{q}|_T \in \mathbf{Nd}^1(T), \ T \in \mathcal{T}_h(\Omega)\}$$

as the associated curl-conforming edge element space. Assuming $\mathbf{g}_{1,h} \in \mathbf{L}^2(\Gamma_1)$ and $\mathbf{g}_{2,h} \in \mathbf{L}^2(\Gamma_2)$ to be appropriately chosen approximations of $\mathbf{g}_1$ and $\mathbf{g}_2$, we set

$$\mathbf{V}_h := \{\mathbf{q}_h \in \mathbf{Nd}^1(\Omega; \mathcal{T}_h(\Omega)) \mid (\nu \times \mathbf{q}_h) \times \nu|_{\Gamma_1} = \mathbf{g}_{1,h}\} \quad , \quad \mathbf{V}_{h,0} := \mathbf{V}_h \cap \mathbf{V_0},$$

and consider the following edge element approximation of (4.5): Find $\mathbf{E}_h \in \mathbf{V}_h$ such that

$$a_{h,\Omega}(\mathbf{E}_h, \mathbf{q}_h) + b_{h,\Gamma_2}(\mathbf{E}_h, \mathbf{q}_h) = \ell_h(\mathbf{q}_h) \quad , \quad \mathbf{q}_h \in \mathbf{V}_{h,0}. \tag{4.6}$$

Here, the sequilinear forms $a_{h,\Omega}(\cdot, \cdot), b_{h,\Gamma_2}(\cdot, \cdot)$, and the functional $\ell_h(\cdot)$ are given by

$$a_{h,\Omega}(\mathbf{E}_h, \mathbf{q}_h) := \sum_{T \in \mathcal{T}_h(\Omega)} \int_T \left(\mu_r^{-1}(\nabla \times \mathbf{E}_h) \cdot (\nabla \times \mathbf{q}_h) - (\kappa^2 \varepsilon_r + i\omega\sigma)\mathbf{E}_h \cdot \mathbf{q}_h \right) d\mathbf{x},$$

$$b_{h,\Gamma_2}(\mathbf{E}_h, \mathbf{q}_h) := \sum_{F \in \mathcal{F}_h(\Gamma_2)} \int_F i\kappa\lambda\nu \times \mathbf{E}_h \cdot ((\nu \times \mathbf{q}_h) \times \nu) \, d\tau,$$

$$\ell_h(\mathbf{q}_h) := \sum_{F \in \mathcal{F}_h(\Gamma_2)} \int_F \mathbf{g}_{h,2} \cdot ((\nu \times \mathbf{q}_h) \times \nu) \, d\tau.$$

We have solved (4.6) by local multigrid with local Hiptmair-Gauss-Seidel smoothing (V-cycle, one pre- and one post-smoothing step) using a hierarchy of four tetrahedral meshes ($L = 3$) created by local refinement on the basis of weighted residual-type a posteriori error estimators. The error estimator consists of weighted element residuals

$$\eta_{T,1}^2 := \alpha_{T,1} \, h_T^2 \, \|\nabla \times (\mu_r^{-1} \nabla \times \mathbf{E}_h) - \kappa^2 \varepsilon_r \mathbf{E}_h\|_{0,T}^2 \quad , \quad T \in \mathcal{T}_h(\Omega),$$

$$\eta_{T,2}^2 := \alpha_{T,2} \, h_T^2 \, \|\kappa^2 \, \nabla \cdot \varepsilon_r \mathbf{E}_h\|_{0,T}^2 \quad , \quad T \in \mathcal{T}_h(\Omega),$$

and weighted face residuals

$$\eta_{F,1}^2 := \alpha_{F,1} \, h_F \, \||\nu \times (\mu_r^{-1} \, \nabla \times \mathbf{E}_h]_F\|_{0,F}^2 \quad , \quad F \in \mathcal{F}_h(\Omega),$$

$$\eta_{F,2}^2 := \alpha_{F,2} \, h_F \, \|\kappa^2 \, [\nu \cdot (\varepsilon_r \mathbf{E}_h)]\|_{0,F}^2 \quad , \quad F \in \mathcal{F}_h(\Omega),$$

$$\eta_{F,3}^2 := \alpha_{F,3} \, h_F \, \|\mathbf{g}_{1,h} - (\nu \times \mathbf{E}_h) \times \nu\|_{0,F}^2 \quad , \quad F \in \mathcal{F}_h(\Gamma_1),$$

$$\eta_{F,4}^2 := \alpha_{F,4} \, h_F \, \|\mathbf{g}_{2,h} - \left(\nu \times (\mu_r^{-1} \nabla \times \mathbf{E}_h) - i\kappa \, \lambda \, \nu \times \mathbf{E}_h\right)\|_{0,F}^2 \quad , \quad F \in \mathcal{F}_h(\Gamma_2),$$

where $[\cdot, \cdot]_F$ stands for the jump across interior faces. We have chosen different weights $\alpha_{T,k}, 1 \leq k \leq 2$, and $\alpha_{F,k}, 1 \leq k \leq 4$, for the element and face residuals in the regions around the transmitter antenna and the receiver antennas to ensure a proper resolution, since the electric field is significantly smaller in the vicinity of the receiver antennas. The initial coarse triangulation of the computational domain is shown in Figure (**4**). Figure (**5**) shows the adaptively refined mesh on the metallic mandrel around the coils of the transmitter antenna (left) and in the region around the aperture (right). We observe a pronounced local refinement in these regions. Figure (**6**) displays the computed electric field $\mathbf{E}_h$ (left) and the computed magnetic induction $\mathbf{B}_h$ (right) in a vicinity of the transmitter antenna along with the adaptively refined mesh. The fields are restricted to the recess around the coils and, as expected, rapidly decay off the transmitter antenna.

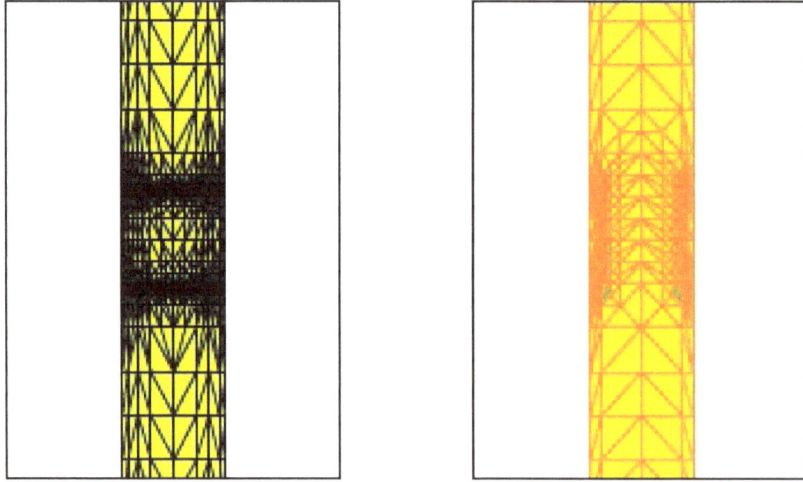

Figure (**5**): Adaptively refined mesh on the mandrel around the coils of the transmitter antenna (l.) and around the aperture (r.)

5 Numerical Simulation of Piezoelectrically Actuated Surface Acoustic Waves (SAW)

Piezoelectric materials are able to generate an electric field in response to an applied mechanical stress which is called the direct piezoelectric effect. The reverse piezoelectric effect is the generation of a mechanical stress and strain under the influence of an applied electric field. The origin of both effects is related to an asymmetry in the unit cell of a piezoelectric crystal which causes a change in the polarization density. It can be observed only in materials with a polar axis (cf., e.g., [55, 56]). Here,

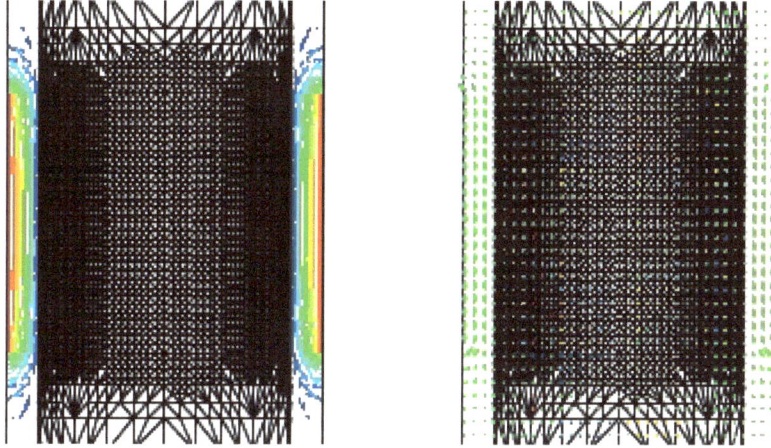

Figure **(6)**: Computed electric field (l.) and magnetic induction (r.) in a vicinity of the transmitter antenna

we are interested in the simulation of piezoelectrically actuated surface acoustic waves [57, 58] with applications in signal processing [59–62] and life sciences [63–67].

In piezoelectric materials, the stress tensor σ depends linearly on the electric field $\mathbf{E}$ according to a generalized Hooke's law

$$\sigma(\mathbf{u}, \mathbf{E}) = \mathbf{c}\varepsilon(\mathbf{u}) - \mathbf{e}\mathbf{E}. \qquad (5.1)$$

Here, $\mathbf{u}$ denotes the mechanical displacement vector and $\varepsilon(\mathbf{u}) := (\nabla\mathbf{u} + (\nabla\mathbf{u})^T)/2$ stands for the linearized strain tensor. Moreover, $\mathbf{c}$ and $\mathbf{e}$ are the symmetric fourth order elasticity tensor and the symmetric third order piezoelectric tensor. Since the frequency of the applied electromagnetic wave is small compared to the frequency of the generated acoustic wave, a coupling can be neglected. Moreover, the electric field is irrotational. Consequently, it can be expressed as the gradient of an electric potential Φ according to $\mathbf{E} = -\nabla\Phi$. Since piezoelectric materials are nearly perfect insulators, the only remaining quantity of interest in Maxwell's equations is the dielectric displacement $\mathbf{D}$ which is related to the electric field by the constitutive equation

$$\mathbf{D} = \varepsilon\mathbf{E} + \mathbf{P}, \qquad (5.2)$$

where ε is the permittivity of the material and $\mathbf{P}$ stands for the polarization. In piezoelectric materials, the polarization $\mathbf{P}$ is linear, i.e., there holds

$$\mathbf{P} = \mathbf{e}\varepsilon(\mathbf{u}). \qquad (5.3)$$

We assume that the piezoelectric material with density $\rho > 0$ occupies some rectangular domain Ω with boundary $\Gamma = \partial\Omega$ and exterior unit normal ν such that

$$\Gamma = \overline{\Gamma}_{E,D} \cup \overline{\Gamma}_{E,N} \quad , \quad \Gamma_{E,D} \cap \Gamma_{E,N} = \emptyset,$$
$$\Gamma = \overline{\Gamma}_{p,D} \cup \overline{\Gamma}_{p,N} \quad , \quad \Gamma_{p,D} \cap \Gamma_{p,N} = \emptyset,$$

where $\Gamma_{E,D}$ is a rectangular subdomain of the upper boundary of Γ and $\Gamma_{E,N} := \Gamma \setminus \overline{\Gamma}_{E,D}$. Given boundary data $\Phi_{E,D}$ on $\Gamma_{E,D}$, the pair $(\mathbf{u}, \Phi)$ satisfies the following initial-boundary value problem for the piezoelectric equations

$$\rho \frac{\partial^2 \mathbf{u}}{\partial t^2} - \nabla \cdot \sigma(\mathbf{u}, \mathbf{E}) = \mathbf{0} \quad \text{in } Q := \Omega \times (0, T) , \tag{5.4a}$$

$$\nabla \cdot \mathbf{D}(\mathbf{u}, \mathbf{E}) = 0 \quad \text{in } Q , \tag{5.4b}$$

$$\mathbf{u} = 0 \quad \text{on } \Gamma_{p,D} , \quad v \cdot \sigma = \sigma_v \text{ on } \Gamma_{p,N} , \tag{5.4c}$$

$$\Phi = \Phi_{E,D} \quad \text{on } \Gamma_{E,D} , \quad v \cdot \mathbf{D} = D_v \text{ on } \Gamma_{E,N} , \tag{5.4d}$$

$$\mathbf{u}(\cdot, 0) = 0, \quad \frac{\partial \mathbf{u}}{\partial t}(\cdot, 0) = 0 \quad \text{in } \Omega . \tag{5.4e}$$

These equations have to be completed by the constitutive equations (5.1),(5.2) and (5.3). Assuming time periodic excitations $\Phi_{E,D}(\cdot, t) = \text{Re}\left(\hat{\Phi}_{E,D} \exp(-i\omega t)\right)$ such that $\hat{\Phi}_{E,D} \in H^{1/2}(\Gamma_{E,D})$, we are looking for time harmonic solutions

$$\mathbf{u}(\cdot, t) = \text{Re}(\mathbf{u}(\cdot) \exp(-i\omega t)) \quad , \quad \Phi(\cdot, t) = \text{Re}(\Phi(\cdot) \exp(-i\omega t)) .$$

This leads to a saddle point problem for a Helmholtz-type equation which in its weak form amounts to the computation of $(\mathbf{u}, \Phi) \in \mathbf{V} \times W$, where $\mathbf{V}_0 := H^1_{0, \Gamma_{p,D}}(\Omega)^3$ and $W := \{\varphi \in H^1(\Omega) \mid \varphi_{\Gamma_{E,D}} = \hat{\Phi}_{E,D}\}$, such that for all $\mathbf{v} \in \mathbf{V}$ and $\psi \in W_0 := H^1_{0, \Gamma_{E,D}}(\Omega)$

$$a(\mathbf{u}, \mathbf{v}) + b(\Phi, \mathbf{v}) - \omega^2 \rho (\mathbf{u}, \mathbf{v})_{0,\Omega} = \ell_1(\mathbf{v}) , \tag{5.5a}$$

$$b(\psi, \mathbf{u}) - c(\Phi, \psi) = \ell_2(\psi) . \tag{5.5b}$$

Here,

$$H^1_{0, \Gamma_{p,D}}(\Omega)^3 := \{\mathbf{v} \in H^1(\Omega)^3 \mid \mathbf{v}|_{\Gamma_{p,D}} = 0\},$$

$$H^1_{0, \Gamma_{E,D}}(\Omega) := \{\psi \in H^1(\Omega) \mid \psi_{\Gamma_{E,D}} = 0\},$$

and the sesquilinear forms $a(\cdot, \cdot), b(\cdot, \cdot), c(\cdot, \cdot)$ and the functionals $\ell_1 \in \mathbf{V}^*, \ell_2 \in W^*$ are given by

$$a(\mathbf{v}, \mathbf{w}) := \int_\Omega \mathbf{c}\varepsilon(\mathbf{v}) : \varepsilon(\bar{\mathbf{w}}) \, d\mathbf{x} \quad , \quad b(\varphi, \mathbf{v}) := \int_\Omega \mathbf{e}\nabla\varphi : \varepsilon(\bar{\mathbf{v}}) \, d\mathbf{x} ,$$

$$c(\varphi, \psi) := \int_\Omega \varepsilon \nabla\varphi \cdot \nabla\bar{\psi} \, d\mathbf{x} ,$$

$$\ell_1(\mathbf{v}) := \langle \sigma_{\mathbf{n}_1}, \mathbf{v} \rangle_{p,N} \quad , \quad \ell_2(\psi) := \langle D_{\mathbf{n}_1}, \psi \rangle_{E,N} ,$$

with $\langle \cdot, \cdot \rangle_{p,N}, \langle \cdot, \cdot \rangle_{E,N}$ denoting the dual pairings between the associated trace spaces and their dual spaces, respectively.

The saddle point problem (5.5a),(5.5b) satisfies the assumptions of Theorem 7.2 of section 3. Hence, if ω is not an eigenvalue of the associated eigenvalue problem, there exists a unique solution $(\mathbf{u}, \Phi) \in \mathbf{V} \times W$.

We have performed numerical simulations of SAWs for plates of length L, width W, and height H such that $\Omega_1 := (0, L) \times (0, W) \times (0, H)$ with $\Gamma_{p,D} := [0, L] \times [0, W] \times \{0\}$. As the piezoelectric material we have assumed Lithiumniobate ($LiNbO_3$). Table 5.1 contains the elasticity tensor $\mathbf{c}$, the piezoelectric tensor ε, the electric permittivity tensor ε, and the density ρ_p of this material.

The IDT has been positioned at the top of the plate, i.e., $\Gamma_{E,D} := [L_1, L_2] \times [W_1, W_2] \times \{L\}$, and has been assumed to operate at a frequency $\omega/(2\pi) = 100 MHz$ thus generating SAWs of wavelength $\lambda = 40\mu m$. In order to control the finite element error appropriately, following [68] we have chosen

Table **1**: Piezoelectric material moduli (Lithiumniobate $LiNbO_3$)

Elast. tensor $10^{10} \frac{N}{m^2}$	$c_{11} = c_{22}$ 20.3	c_{12} 5.3	$c_{13} = c_{23}$ 7.5	$c_{14} = -c_{24} = c_{56}$ 0.9	c_{33} 24.5	$c_{44} = c_{55}$ 6.0	c_{66} 7.5
Piezoel. tensor $\frac{C}{m^2}$	$e_{15} = e_{24}$ 3.7		$e_{22} = -e_{21} = -e_{16}$ 2.5		$e_{31} = e_{32}$ 0.1		e_{33} 1.3
Permitt. tensor $10^{-12} \frac{F}{m}$	$\varepsilon_{11} = \varepsilon_{22}$ 749.0	ε_{33} 253.2	**Density** $10^3 \frac{kg}{m^3}$		ρ_p 4.63		

an initial mesh of mesh length $h \lesssim \sqrt{\lambda^3}$.

For a plate of length $L = 1.2\,mm$, width $W = 0.6\,mm$, height $H = 0.6\,mm$, and $\Gamma_{E,D} := [0.2, 0.4] \times [0.1, 0.5] \times \{1.2\}$, Figure **(7)** (left) shows the computed amplitudes of the electric potential wave for the longitudinal section $[0, 1.2] \times \{0.3\} \times [0, 0.6]$.

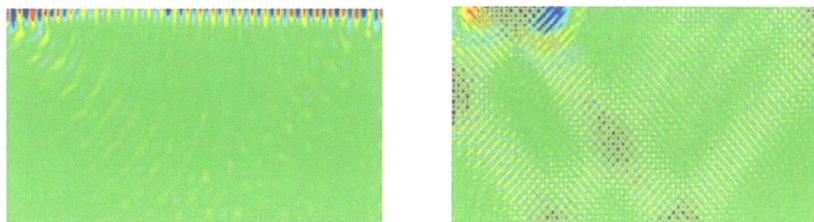

Figure **(7)**: Amplitudes of a surface acoustic wave (100 MHz) (l.) and bulk wave (200 MHz) (r.)

As can be clearly seen, the SAWs are strictly confined to the surface of the piezoelectric material with a penetration depth of approximately one wavelength as it is required for their application in nano-pumps for SAW driven microfluidic biochips. The SAW velocity is $4.0 \cdot 10^3\,m/s$. In contrast to this, Figure **(7)** (right) displays a typical bulk wave generated by an IDT operating at a frequency of 200 MHz which ia a wave configuration useful for applications in telecommunications (cell phones).

For a simplified test case from [65], Table 5.2 and Table 5.3 reflect the convergence histories of the iterative schemes without and with the BPX-type preconditioner.

Acknowledgments

The work has been supported by the NSF under Grant No. DMS-0810176. The work of the second and third author has been supported by by the National Basic Research Program of China (Grant No. 2005CB321701) and the National Natural Science Foundation of China (Grant No. 10731060)

Table **2**: Number of iterations and CPU-time (in seconds) for SC-CG and BICGSTAB/GMRES without using a preconditioner

Level	SC–CG		BICGSTAB		GMRES	
	time	iter	time	iter	time	iter
3	0.15	74	0.10	65	0.14	17
4	1.4	148	0.75	137	1.7	56
5	29	311	7.6	324	32	206
6	440	872	75	678	530	758

Table **3**: Number of iterations and CPU-time (in seconds) for SC-PCG and BICGSTAB/GMRES with a block-diagonal preconditioner

Level	SC–PCG		PBICGSTAB		PGMRES	
	time	iter	time	iter	time	iter
5	2.5	48	1.1	33	1.2	6
6	12	52	5.2	39	5.9	7
7	70	55	23	41	25	7
8	290	57	92	44	100	8

Bibliography

[1] Bramble, JH. Multigrid methods. Boston: Pitman 1993.

[2] Hackbusch, W. Multigrid methods and applications. Berlin-Heidelberg-New York: Springer 1985.

[3] Oswald, P. Multilevel finite element approximation: theory and applications. Stuttgart: Teubner 1994.

[4] Quarteroni, A, Valli, A. Domain decomposition methods for partial differential equations. Oxford: Clarendon Press 1999.

[5] Smith, BF, Björstad, PE, Gropp, WD. Domain decomposition methods. Cambridge: Cambridge University Press 1996.

[6] Toselli, A, Widlund, O. Domain decomposition methods - algorithms and theory. Berlin-Heidelberg-New York: Springer 2005.

[7] Trottenberg, U, Oosterlee, K, Schüller, A. Multigrid. New York: Academic Press 2000.

[8] Braess, D, Hackbusch, W. A new convergence proof for the multigrid method including the V-cycle. SIAM J Numer Anal 1983; 36:967-975.

[9] Bramble, JH, Pasciak, JE. New estimates for multigrid algorithms including the V-cycle. Math Comp 1993; 60:447-471.

[10] Brenner, SC. Convergence of the multigrid V-cycle algorithms for the second order boundary value problems without full elliptic regularity. Math Comp 2002; 71:507-525.

[11] Yserentant, H. On the multi-level splitting of finite element spaces. Numer Math 1986; 49:379-412.

[12] Yserentant, H. Two preconditioners based on the multi-level splitting of finite element spaces. Numer Math 1990; 58:163-184.

[13] Yserentant, H. Old and new convergence proofs for multigrid methods. Acta Numerica 1993; 2:285-326.

[14] Xu, J. Iterative methods by space decomposition and subspace correction. SIAM Review 1992; 34:581-613.

[15] Ainsworth, M, Oden, JT. A posteriori error estimation in finite element analysis. New York: John Wiley & Sons 2000.

[16] Bangerth, W, Rannacher, R. Adaptive finite element methods for differential equations. Lectures in Mathematics. ETH-Zürich. Basel: Birkhäuser 2003.

[17] Eriksson, K, Estep, D, Hansbo, P, Johnson, C. Computational differential equations. Cambridge: Cambridge University Press 1996.

[18] Neittaanmäki, P, Repin, S. Reliable methods for mathematical modelling. Error control and a posteriori estimates. New York: Elsevier 2004.

[19] Verfürth, R. A review of a posteriori error estimation and adaptive mesh-refinement techniques. Chichester: Wiley-Teubner 1996.

[20] Morin, P, Nochetto, RH, Siebert, KG. Convergence of adaptive finite element methods. SIAM Review 2002; 44:631-658.

[21] Binev, P, Dahmen, W, DeVore, R. Adaptive finite element methods with convergence rates. Numer Math 2004; 97:219-268.

[22] Cascon, JM, Kreuzer, C, Nochetto, RH, Siebert, KG. Quasi-optimal convergence rate for an adaptive finite element method. SIAM J Numer Anal 2008; 46:2524-2550.

[23] Stevenson, R. Optimality of a standard adaptive finite element method. Foundations of Computional Mathematics 2007; 2:245-269.

[24] Carstensen, C, Hoppe, RHW. Convergence analysis of an adaptive edge finite element method for the 2d eddy current equations. J. Numer Math 2005; 13:19-32.

[25] Hoppe, RHW, Schöberl, J. Convergence of adaptive edge element methods for the 3D eddy currents equations. J Comp Math 2009; 27:657-676.

[26] Zhong, L, Chen, L, Shu, S, Wittum, G, Xu, J. Quasi-optimal convergence of adaptive edge finite element methods for three dimensional indefinite time-harmonic Maxwell's equations. Report. Department of Mathematics, University of California at Irvine 2010.

[27] L. Zhong, L. Chen, and J. Xu, Convergence of adaptive edge finite element methods for H(curl)-elliptic problems. Numer Lin Algebra Appl **17**, 415-432, 2009.

[28] McCormick, S. Multilevel Adaptive methods for partial differential equations. Frontiers in Applied Math. Vol. 6. Philadelphia: SIAM, 1989.

[29] Widlund, O. Optimal iterative refinement methods. In: Chan, T, Glowinski, R, Périaux, J, Widlund, eds., Domain Decomposition Methods. Philadelphia: SIAM 1989; pp. 114-125.

[30] Bai, D, Brandt, A. Local mesh refinement multilevel techniques. SIAM J Sci Stat Comput 1987; 8:109-134.

[31] Bramble, JH, Pasciak, JE, Wang, J, Xu, J. Convergence estimates for product iterative methods with applications to domain decomposition. Math Comp 1991; 57:23-45.

[32] Ainsworth, M, McLean, W. Multilevel diagonal scaling preconditioners for boundary element equations on locally refined meshes. Numer Math 2003; 93:387-413.

[33] Aksoylu, B, Bond, S, Holst, M. An odyssee into local refinement and multilevel preconditioning III: implementation and numerical experiments. SIAM J Sci Comput 2003; 25:478-498.

[34] Aksoylu B, Holst, M. Optimality of multilevel preconditioners for local mesh refinement in three dimensions. SIAM J Numer Anal 2006; 44:1005-1025.

[35] Dahmen, W, Kunoth, A. Multilevel preconditioning. Numer Math 1992; 63:315-344.

[36] Rüde, U. Fully adaptive multigrid methods. SIAM J Numer Anal 1993; 30:230-248.

[37] Wu, HJ, Chen, ZM. Uniform convergence of multigrid V-cycle on adaptively refined finite element meshes for second order elliptic problems. Science in China 2006; 39:1405-1429.

[38] Xu, X, Chen, H, Hoppe, RHW. Optimality of local multilevel methods on adaptively refined meshes for elliptic boundary value problems. J. Numer Math 2010; 18:59-90.

[39] Ciarlet, PG. The finite element method for elliptic problems. Philadelphia: SIAM 2002.

[40] Monk, P. Finite element methods for Maxwell's equations. Oxford: Clarendon Press 2003.

[41] Nédélec, JC. Mixed finite elements in $\mathbb{R}^3$. Numer Math 1980; 35:315-341.

[42] Beck, R, Hiptmair, R, Hoppe, RHW, Wohlmuth, B. Residual based a posteriori error estimators for eddy current computation. M^{2}AN Math Modeling and Numer Anal 2000; 34:159-182.

[43] Hiptmair, R. Multigrid method for Maxwell's equations. SIAM J Numer Anal 1998; 36:204-225.

[44] Hiptmair, R. Finite elements in computational electromagnetism. Acta Numerica 2002; 11:237-339.

[45] Arnold, D, Falk, R, Winther, R. Multigrid in H(div) and H(curl). Numer Math 2000; 85:197-218.

[46] Brezzi, F, Fortin, M. Mixed and hybrid finite element methods. Berlin-Heidelberg-New York: Springer 1991.

[47] Zeidler, E. Nonlinear functional analysis and its applications. II/A: Linear monotone operators. Berlin-Heidelberg-New York: Springer 1990.

[48] Axelsson, O. Iterative solution methods. Cambridge: Cambridge University Press 1996.

[49] Saad, Y. Iterative methods for sparse linear systems. Cambridge: Cambridge University Press 2003.

[50] Klawonn, A. An optimal preconditioner for a class of saddle point problems with a penalty term. SIAM J Sci Comput 1998; 19:540–552.

[51] Axelsson, O, Kolotilina, LY, Eds. Preconditioned conjugate gradient methods. Berlin-Heidelberg-New York: Springer 2008.

[52] Bramble, JH, Pasciak, JE, Xu, J.Parallel multilevel preconditioners. Math Comp 1990; 55:1-22.

[53] Schlumberger Oilfield Services. Private communication. Sugarland, TX, 2010.

[54] Buffa,A, Costabel, M, Sheen, D. On traces for $H(curl, \Omega)$ in Lipschitz domains. J Math Anal Appl 2002; 276:845-867.

[55] Eringen, AC, Maugin, GA. Electrodynamics of continua I. Foundations and solid media. Berlin-Heidelberg-New York: Springer 1990.

[56] Maugin, GA. Continuum mechanics of electromagnetic solids. Amsterdam: North-Holland 1987.

[57] Gualtieri, JG, Kosinski, JA, Ballato, A. Piezoelectric materials for acoustic wave applications. IEEE Trans. Ultrasonics, Ferroelectrics, and Frequency Control 1994; 41:53-59.

[58] Günter, P, Huignard, JP, Eds. Photorefractive materials and their applications 1. Basic effects. Berlin-Heidelberg-New York: Springer 2006.

[59] Eason, RW, Miller, A. Nonlinear optics in signal processing. London: Chapman & Hall 1993.

[60] Feldmann, M, Hénaff, J. Surface acoustic waves for signal processing. Boston: Artech House 1989.

[61] Kino, GS. Acoustic waves: devices, imaging, and analog signal processing. Englewood Cliffs: Prentice-Hall 1987.

[62] Morgan, DP. Surface-wave devices for signal processing. Amsterdam: Elsevier 1991.

[63] Antil, H, Gantner, A, Hoppe, RHW, Köster, D, Siebert, KG, Wixforth, A. Modeling and simulation of piezoelectrically agitated acoustic streaming on microfluidic biochips. In: Langer, U, *et al.* , Eds. Proceedings 17th Int. Conf. on Domain Decomposition Methods, Lecture Notes in Computational Science and Engineering Vol. 60. Berlin Heidelberg-New York: Springer 2007; pp. 305-312.

[64] Antil, H, Glowinski, R, Hoppe, RHW, Linsenmann, C, Pan, TW, Wixforth, A. Modeling, simulation, and optimization of surface acoustic wave driven microfluidic biochips. J Comp Math 2010; 28:149-169.

[65] Gantner, A, Hoppe, RHW, Köster, D, Siebert, KG, Wixforth, A. Numerical simulation of piezo-electrically agitated surface acoustic waves on microfluidic biochips. Comp Visual Sci 2007; 10:145-161.

[66] Wixforth, A. Acoustically driven programmable microfluidics for biological and chemical applications. JALA 2006; 11:399-405.

[67] Wixforth, A, Strobl, C, Gauer, C, Toegl, A, Scriba, J, Guttenberg, Z. Acoustic manipulation of small droplets. Anal Bioanal Chem 2004; 379:982-991.

[68] Ihlenburg, F. Finite element analysis of acoustic scattering. New York: Springer 1998.

Index

www.ingramcontent.com/pod-product-compliance
Lightning Source LLC
Chambersburg PA
CBHW041712210326
41598CB00007B/618